Big Science and Research Infrastructures in Europe

Big Science and Research Infrastructures in Europe

Edited by

Katharina C. Cramer

Lund University, Sweden

Olof Hallonsten

Lund University, Sweden

Cheltenham, UK • Northampton, MA, USA

Published by
Edward Elgar Publishing Limited
The Lypiatts
15 Lansdown Road
Cheltenham
Glos GL50 2JA
UK

Edward Elgar Publishing, Inc.
William Pratt House
9 Dewey Court
Northampton
Massachusetts 01060
USA

A catalogue record for this book
is available from the British Library

Library of Congress Control Number: 2020940515

This book is available electronically in the **Elgar**online
Social and Political Science subject collection
http://dx.doi.org/10.4337/9781839100017

ISBN 978 1 83910 000 0 (cased)
ISBN 978 1 83910 001 7 (eBook)

Printed by CPI Group (UK) Ltd, Croydon CR0 4YY

Contents

Tables

Contributors

Isabel K. Bolliger is a PhD researcher at the Swiss Graduate School of Public Administration. Her research focuses on decision-making processes for priority setting and funding of large-scale research infrastructures. Since her graduation from the University of Zurich in 2014 she has been working in different institutions in the research policy sector. She was a consortium member of the Horizon 2020 project InRoad and contributed also to other activities related to policy formulation for research infrastructures.

August Collsiöö is a doctoral student in cognitive psychology at Uppsala University, Sweden. His research is focused on judgment and decision making, especially targeting how presenting information in a qualitative or quantitative manner impacts judgments and decisions. He additionally performs research related to analytic and intuitive thought. August was previously employed as a research and evaluation consultant at Oxford Research, a specialized knowledge company, where the work found in this book was carried out.

Katharina C. Cramer is a historian specialized in the history and politics of science and technology and the power of knowledge and innovation in the twentieth and twenty-first centuries. She studied political science, sociology and history at the universities in Bonn, Germany, Istanbul, Turkey and Luxembourg and completed her doctorate in history at the University of Konstanz, Germany in 2018.

Beatrice D'Ippolito is Associate Professor of Strategic Management at the University of York Management School, United Kingdom. Her research focuses on the emergence and development of mechanisms of knowledge creation and diffusion that connect the micro level of firm dynamics with the meso level of industry evolution. She has conducted extensive qualitative and archival research across different empirical domains, that are, design, science collaborations, renewable energies and creative industries. Her research has been published in top-tier academic journals such as *Research Policy*, *Technovation* and *Industry and Innovation.*

Hjalmar Eriksson is currently Project Manager for external relations at Stockholm University. He was previously a research and evaluation consultant at Oxford Research, a specialized knowledge company, where he delivered services within knowledge and innovation systems and sustainable development to the public sector, e.g. to the Swedish Radiation Safety Authority. His professional interests center on the evolution of society through the utility and impact of knowledge and research.

Thomas Franssen is a researcher at the Centre of Science and Technology Studies, Leiden University, The Netherlands. He has a PhD in sociology (University of Amsterdam, 2015) and works on the boundary of sociology, science and technology studies and valuation studies. His current research analyzes changes in research governance and the epistemic effects of such changes in different scientific domains. His chapter is part of a project that studies how digitalization of the research environment changes interdependence relations in the humanities.

Alexandra Griffiths is a graduate of the Swiss Graduate School of Public Administration. She wrote her dissertation on the influence of ESFRI on national roadmapping processes. She was also a consortium member of the Horizon 2020 InRoad project and co-authored a report on best practices and trends in research infrastructure roadmapping in 2018. She obtained her master of arts in public management and policy in 2019.

Olof Hallonsten is a sociologist of science, specialized in studies of the politics and organization of science and the interfaces between science and other parts of society. He teaches organization studies at the Lund University School of Economics and Management, and has previously worked at the universities of Gothenburg, Sweden, and Wuppertal and Bamberg, Germany. His works include several journal articles on Big Science, and on the politics, organization and funding of science in various national contexts.

Jean-Christophe Mauduit is a lecturer in science diplomacy at the Department of Science, Technology, Engineering in Public Policy, University College London. He was initially trained in physics and holds a PhD in astronomy. He then worked as a researcher on European Space Agency and NASA satellite missions and as a project officer for the International Astronomical Union. He recently obtained an MA in law and diplomacy and was a research scholar at the American Association for the Advancement of Science.

Maria Moskovko is a doctoral student in research policy at Lund University, Sweden. She is interested in how contemporary European Research Infrastructure projects come into being as distinct legal entities and what role (if any) the European Union plays in those processes and how collaborative Research Infrastructures emerge from scientific ideas and get set up as organizations under particular legal arrangements. Prior to her research pursuits at

Lund University, she worked at the Council of Europe and American Councils for International Education.

Nicolas Rüffin is Research Fellow of the President's Project Group at the WZB Berlin Social Science Center, Germany. He is currently doing his doctorate at the University of Kassel, Germany. He received a master's degree in science studies from Humboldt-University in Berlin and a bachelor's degree in business psychology from the University of Bochum, Germany.

Charles-Clemens Rüling is Full Professor of Organization and Management Theory and Associate Dean for Research at Grenoble Ecole de Management, France. He holds a postdoctoral research degree in management, a doctorate in social and economic sciences and master degrees in sociology and business administration. He is an associate editor of *Long Range Planning* and a permanent member of the IREGE research lab, Université Savoie Mont Blanc, France. Charles-Clemens Rüling teaches organization theory and research methods, and his main research focuses on institutional dynamics and field configuring events.

Inga Ulnicane currently undertakes research on governance of emerging technologies, artificial intelligence and dual use research at De Montfort University, United Kingdom. She has more than ten years of international and interdisciplinary research, teaching and engagement experience in the areas of science, technology and innovation policy and governance. She has published on topics such as international research collaboration, the European Research Area and grand societal challenges. She has also undertaken commissioned studies for the European Commission and the European Parliament.

Andrew Williams is External Relations Officer at the European Southern Observatory and is responsible for strategic government relations. He worked for over a decade as Senior Policy Advisor for the North Atlantic Treaty Organization and as a physicist for the government of the United Kingdom. He holds a degree in physics and a doctorate in public policy. He has authored 15 peer-reviewed articles and chapters in defense policy, administration and political science topics, and now researches Big Science organizations and policy.

Abbreviations

AWTI	The Netherlands Advisory Council for Science, Technology and Innovation (Adviesraad voor Wetenschap, Technologie en Innovatie)
BBMRI	Biobanking and Biomolecular Resources Research Infrastructure
CERIC	Central European Research Infrastructure Consortium
CERN	European Organization for Nuclear Research (Conseil Européen pour la Recherche Nucleaire)
CESSDA	Consortium of European Social Science Data Archives
CLARIAH	Common Lab Research Infrastructure for the Arts and Humanities
CLARIN	Common Language Resources and Technology Infrastructure
DARIAH	Digital Research Infrastructure for the Arts and Humanities
DESY	German Electron Synchrotron (Deutsches Elektronen-Synchrotron)
EATRIS	European Infrastructure for Translational Medicine
EC	European Community
ECCSEL	European Carbon Dioxide Capture and Storage Laboratory Infrastructure
ECRI	European Conference on Research Infrastructures
ECRIN	European Clinical Research Infrastructure Network
EEC	European Economic Community
EEIG	European Economic Interest Grouping
EHPG	Enlarged Halden Program Group
EIROforum	European International Research Organisation Forum
ELDO	European Space Vehicle Launcher Development
ELI	Extreme Light Infrastructure

ELIXIR	European Life-Science Infrastructure for Biological Information
ELT	Extremely Large Telescope
EMBL	European Molecular Biology Laboratory
EMSO	European Multidisciplinary Seafloor and Water Column Observatory
ERA	European Research Area
ERIC	European Research Infrastructure Consortium
ESA	European Space Agency
ESFRI	European Strategy Forum on Research Infrastructures
ESO	European Southern Observatory
ESRF	European Synchrotron Radiation Facility
ESRO	European Space Research Organisation
ESS	European Spallation Source
ETW	European Transonic Wind Tunnel
EU	European Union
European XFEL	European X-Ray Free-Electron Laser
FAIR	Facility for Antiproton and Ion Research
FET	future emerging technologies
FP	Framework Programme for Research and Technological Development
GSF	Global Science Forum
GSO	Group of Senior Officials
HAMMLAB	Halden Man-Machine Laboratory
HBP	Human Brain Project
HBWR	Halden Boiling Water Reactor
HPG	Halden Program Group
HRP	Halden Reactor Project
HWR	Halden Working Reports
IAU	International Astronomical Union
ICOS	Integrated Carbon Observation System
ICT	Information and Communication Technology

IFE	Norwegian Institute for Energy Technology (Institutt for Energiteknikk)
ILL	Institut Laue-Langevin
Infrafrontier	The European Infrastructure for Phenotyping and Archiving of Model Mammalian Genomes
INSTRUCT	Integrated Structural Biology Infrastructure
IO	international organization
ITER	International Thermonuclear Experimental Reactor
JHR	Jules Horowitz Reactor
JIVE	Joint Institute for Very Long Baseline Interferometry
KNAW	Royal Netherlands Academy of Arts and Science (Koninklijke Nederlandse Akademie van Wetenschappen)
KSU	Nuclear Safety and Training (Kärnkraftsäkerhet och Utbildning)
LIFEWATCH	Science and Technology Infrastructure for Biodiversity Data and Observatories
MTO	Man-Technology-Organization
NSF	National Science Foundation
NWO	The Netherlands Science Funding Organization (Nederlands Organisatie voor Wetenschappelijk Onderzoek)
OECD	Organisation for Economic Co-operation and Development
OSP	open skies policy
PRACE	Partnership for Advanced Computing in Europe
R&D	research and development
RAMIRI	Realising and Managing International Research Infrastructures
RI	Research Infrastructure
RWG	Roadmap Working Group
SEA	Single European Act
SHARE	Survey of Health, Ageing and Retirement in Europe
SKA	Square Kilometre Array
SLAC	Stanford Linear Accelerator Center

SPIRAL2	Système de Production d'Ions Radioactifs en Ligne de 2e génération
SSC	Superconducting Supercollider
SSM	Swedish Radiation Safety Authority (Strålsäkerhetsmyndigheten)
STS	Science and Technology Studies
SWG	Strategy Working Group
UK	United Kingdom
US	United States
VAT	value added tax
WoS	Web of Science

1. Big Science and Research Infrastructures in Europe: History and current trends

Katharina C. Cramer, Olof Hallonsten, Isabel K. Bolliger and Alexandra Griffiths

1. INTRODUCTION

This book is about two widely used but scarcely analyzed current phenomena: Big Science and Research Infrastructures (RIs). Both concepts are enormously important in European science, technology and innovation policy, judging by the political "hypes" around them (Hallonsten 2020a) and because of their evident inherent forces for European integration and multilateral cooperation and diplomatic progress (Krige 2003; Papon 2004; Hallonsten 2014; Cramer 2017). On the basis of the current use of the term Research Infrastructures and the historical and current use of the term Big Science, it can be established that the categories of things that the two describe are very wide and partially overlapping. Classic examples of Big Science, like particle accelerators and telescopes, are certainly infrastructures used for scientific research, and hence in some rudimentary sense RIs. Therefore, the politics and history of the two are inseparable, especially perhaps in a European context where the need and will for collaboration, coordination and integration among a varying number of countries make most costly and advanced facilities and resources for scientific research unavoidably political.

While the term Research Infrastructures seems to represent a more recent phenomenon, the era of Big Science dates back to at least the early Cold War. After World War II, investments in science and technology soared due to the powerful and partly horrendous wartime demonstrations of the potential of science and technology to alter and shape geopolitics, the postwar economic growth as well as the eagerness of governments to invest in national competitiveness, and the advancement of ideals of a modern and rationally planned society. Major, large-scale and goal-oriented programs to develop weapons, energy production, transport systems, and the engineering of new materials for use in military and civilian technologies, were initiated in all industrialized countries. Big Science became both a role model and a deterrent, and its connections with nuclear energy, nuclear weapons and superpower competition between the United States (US) and the Soviet Union tied it closely to the bipolar geopolitical world order (Galison and Hevly 1992; Hallonsten 2016a). Accelerators for particle physics experiments were built on both sides of the Iron Curtain to demonstrate superiority over the other superpower, and research and development (R&D) programs with peaceful and military motivations were institutionally entangled in governmental research laboratories and followed the cycles of growth and decline of each other and the greater geopolitical tides of hostility and détente (Greenberg 1967/99; Hallonsten and Heinze 2012).

Big Science in Europe also evolved very much in accordance with the political logic brought about by the historical developments during the Cold War, including the arms race, but perhaps most importantly, the need to collaborate to secure long-term peace on the continent as well as the desire for restoration of national sovereignty (Krige 2006). In this regard, the launch of the European Organization for Nuclear Research (CERN), a collaborative nuclear and particle physics laboratory, in the 1950s was Europe's first postwar experience in large-scale cooperation (Hermann et al 1987, 1990; Krige 1996). Other examples of Big Science in Europe, extensively studied by historians and sociologists, are the European Southern Observatory (ESO) (Blaauw 1991; Madsen 2012), the Institut Laue-Langevin (ILL) (Pestre 1997; Jacrot 2006; D'Ippolito and Rüling 2019; D'Ippolito and Rüling, ch 11 in this volume), the European Synchrotron Radiation Facility (ESRF) (Cramer 2017, 2020; Simoulin 2012) and the German Electron Synchrotron (DESY) (Heinze et al 2015a, 2015b, 2017). Attention has also been paid to the broader historical, political and macrosociological developments of Big Science in Europe (Krige 2003; Hallonsten 2014, 2016a; Cramer 2020).

In a scholarly meaning, however, Big Science was defined retrospectively. The early use of the concept was by practitioners and commentators who analyzed statistical manifestations of a growing scientific enterprise (Price 1963/86) and warned for the most extreme consequences of team- and program-oriented research, such as bureaucratization at the expense of creativity and academic freedom (Weinberg 1961, 1967). As a conceptual label in the history of science, Big Science got its breakthrough in the early 1990s, initially as an umbrella term for investigations of the particular dynamics of science and technology in the US and Europe during the Cold War. In this context, Capshew and Rader (1992: 22) proposed a differentiation between "big science" and "Big Science", with the latter being "a rhetorical construction" pointing to the particular dynamics of large-scale research following the end of World War II, and the former a generic label used to illustrate the growth of science in size or numbers in the twentieth century.

In this book we deal with Big Science, capitalized. Our interest lies not in the general growth of science in society, but particular forms of science that usually require large-scale instrumentation. We do, however, note that throughout the most recent decades, the concept of Big Science has become commonplace in the history and sociology of science and it has thus also grown remarkably broad and heterogeneous: Nowadays, things typically identified as Big Science are still manifest features of science and innovation systems, although their expressions are less conspicuously "big" and less distinctly connected to military interests and geopolitics (Hallonsten 2016a).

Looking at current European science and innovation policy, especially as represented in the ubiquity of documents, reports and policy memoranda produced by the European Union (EU), the current era in the early twenty-first century seems to be an era of RIs. In the past two to three decades, the EU has undertaken a comprehensive policymaking effort to strengthen the global competitiveness of its common market by support of science, technology and innovation for growth, sustainability and prosperity. RIs are defined by the EU as "facilities, resources and related services that are used by the scientific community to conduct top-level research in their respective fields" (European Council 2009: 4). The concept has become very important as various projects, organizations and installations are branded as RIs in policy documents and EU legislation, and identified as crucially important for the long-term competitiveness of European science and innovation.

Yet, as a concept, Research Infrastructures is very wide and very varied, and there are no definitions available that meet the rudimentary standards of scholarly analysis (Hallonsten 2020a). An intuitive or traditional interpretation of research infrastructures (non-capitalized) could be comparably large and centralized physical and technically advanced resources that are used for experimental research in the natural sciences. This would also mean significant overlap with a traditional or intuitive interpretation of Big Science (capitalized). But the term also has a different meaning and there is hence reason to separate "research infrastructures" and "Research Infrastructures", similarly to what Capshew and Rader (1992) suggested with regard to "Big Science" and "big science" (above). In this regard, it needs to be acknowledged that the concept Research Infrastructures (capitalized) is very much connected to policymaking in Europe, where it ostensibly refers to a rather wide collection of entities that include some typical Big Science but also physically distributed resources for research, such as computing networks or large collections of data or physical objects, with user communities in the social sciences and humanities; or mobile vehicles, like aircrafts for atmospheric testing or icebreaker ships for polar expeditions. The collections of projects and organizations in the many roadmap documents that are produced by European bodies and national agencies in several European countries to coordinate RIs in Europe show a remarkable variety in this regard (see below and Bolliger and Griffiths, ch 5 in this volume), as do the to date 21 organizations that have been granted status as the European Research Infrastructure Consortium (ERIC) (Moskovko, ch 6 in this volume).

There is, hence, a certain overlap between the concepts Big Science and Research Infrastructures (both capitalized), but both concepts can also be given broader interpretations and extend beyond big physical installations, to include mission-oriented research in larger organizations, such as genomics projects, space programs and explorer missions. The two concepts are, however, also connected by history, and more specifically, European history: The recent emergence of RIs as a policy area in Europe follows a longer historical trajectory where Big Science has had a key role, within a framework of (geo)politics of the European continent and its forces of integration and disintegration throughout the post-World War II era (Cramer, ch 3 in this volume). A historical inability of European countries to agree and align their interests in intergovernmental, collaborative large-scale science projects (see Section 2), and a perceived need of the European Economic Community (EEC) and its successor the EU (these acronyms are used alongside each other throughout this chapter, to denote the community of Europe before and after the 1992 Maastricht Treaty) to actively engage in research and innovation policy, evidently produced the current policy "hype" around RIs in Europe, and launched the policy area itself.

2. A BRIEF HISTORY OF BIG SCIENCE AND RESEARCH INFRASTRUCTURES IN EUROPE

As noted, the phenomenon of Big Science dates back to the early Cold War period when science and technology became increasingly important for national power and security policies. The technology- and science-based conduct of war in the early-to-mid-twentieth century and the practical goal-oriented use and development of science and technology within large-scale research projects funded by governments in this era tied these projects closely to military, political and economic interests. Moreover, with the emerging (geo)political tensions between the US and the Soviet Union, the two powers sought to compete also beyond territorial gains and diplomatic successes, and large scientific projects became instruments for long-term security ambitions, and symbols of scientific and technological capability (Greenberg 1967/99; Krige 2006).

Fundamental research in physics, with its clear connection to nuclear energy and weapons development, got a role as essential contributor to both

US-American and Soviet military power during the Cold War. The space programs on both sides of the Iron Curtain also became integrated parts of the so-called Military-Industrial Complex, to which "scientific" could well be added as a third ingredient (Hallonsten 2016a: 43ff.; Giroux 2007). Moreover, at that time, the construction of ever more complex and costly instruments became a decisive factor for success and progress in scientific fields such as ground-based astronomy and nuclear/particle physics. For instance, the ever larger accelerators for nuclear/particle physics demanded ever larger governmental investments (Hoddeson et al 2008; Greenberg 1967/99).

In the immediate post-World War II years, Western Europe was marked by destruction and an unstable political situation. But the agreement on the Schuman plan, the establishment of the European Coal and Steel Community and the Marshall Plan created a political climate in favor of collaboration. The 1957 launch of the Common Market and the European Atomic Energy Community, by the Treaties of Rome, institutionalized intergovernmental collaboration in some specific areas, but it did not include science and technology (beyond nuclear energy) (Guzzetti 1995; Krige and Guzzetti 1997; Tindemans 2009; Hallonsten 2014).

It was not until the 1970s that the EEC expanded its policy areas into the promotion of science and technology. The initiation of industry-oriented programs, to increase competitiveness in specific sectors, was certainly triggered by the economic downturn and the reevaluation of the science and innovation policy doctrines that had been in place since the war (Elzinga and Jamison 1995; Guston 2000). The first Framework Programme for Research and Technological Development in 1984 and the signing of the Single European Act in 1986 constituted further important milestones with regard to the establishment of a more coherent and strategic research policy agenda of the EEC (Guzzetti 1995: 83–6; Papon 2004: 69–70). Yet, although at the end of the 1980s "[t]he mechanics for the implementation of a European research policy were practically put in place" (Papon 2012: 49), active partaking of the European Commission in maintaining and developing a broad research base in Europe began only several decades later.

The establishment of CERN in 1954 by 20 states was Europe's first experience in intergovernmental collaborative large-scale research. Its creation was a particular symbol for a renewed Western (political) unity that should strengthen regional collaboration and political integration (Krige 2003, 2006). A large number of intergovernmental large-scale research collaborations followed in the coming decades: ESO in 1962, the European Space Research Organisation (ESRO) in 1962, the European Space Vehicle Launcher Development (ELDO) in 1964, ILL in 1967, the European Molecular Biology Laboratory (EMBL) in 1973 and ESRF and the European Transonic Wind Tunnel (ETW) in 1988. All these projects were based on intergovernmental agreements signed by a varying number of countries; a mode of cooperation that became known as "variable geometry" or "Europe à la carte" (Papon 2009: 36; Guzzetti 1995; Ulnicane, ch 4 in this volume) and also extended beyond collaborative large research projects, to multilateral coordinating activities such as the European Cooperation in Science and Technology (COST), the European Research Coordination Agency (EUREKA) and high-technology and industry-related joint ventures such as the multinational aerospace corporation Airbus or the launch vehicle Ariane (Roland 1997: 356; Peterson 1997; Papon 2004: 68).

In spite of their formal detachment from common EEC and EU policymaking, large projects in science and technology came to play important roles in European history and the European integration process in the second half of the twentieth century. In the analysis of historians Misa and Schot (2005: 2), "scientific and technical cooperation among national states in Europe proved to be a catalyst for political integration, and a compensation for setbacks in the political area" during the 1960s, as well as an answer to the threat of Soviet and US dominance. The successful establishment of European space efforts, such as the European Space Agency, Ariane and Airbus are also identified as significant for the construction (and consolidation) of European power and a European community: "These large technological projects generated a practical sense of what Europe was about, while producing highly visible outcomes such as Ariane and Airbus that served as symbols for European power and might" (Misa and Schot 2005: 14).

The creation of the European Research Area (ERA) in the early 2000s certainly was a key milestone within a broader policy approach to improve the global role of the EU in science and technology, and the globalized knowledge-based economy. The broader development of EU research policy in the decade preceding the launch of ERA can be interpreted as a shift of focus to *innovation*, which meant a broadening in scope beyond industry-related issues and to a more systemic view with a broader set of actors and institutions included at the receiving end of policymaking (Elera 2006; Hallonsten 2020a). In this context, large-scale research projects, eventually labelled Research Infrastructures in EU policy documents, got increased attention from EU policymakers and administrators. According to EU policy documents published in the first decade of this century, RIs cover a variety of different large-scale research projects from national as well as collaborative efforts, single-sited and distributed projects, or single large instruments, networks and databanks in various disciplines and scientific fields (natural sciences and humanities alike) (see below).

RIs had gotten some attention already in earlier Framework Programmes, such as in the second (in place 1987–91) where 30 million ECU was allocated for facilitating access to RIs. But the real breakthrough for RIs in EU policymaking came with ERA and the 2002 creation of the European Strategy Forum on Research Infrastructures (ESFRI). The early 2000s was a period when Europe experienced especially dire difficulties to agree and push forward collaborative projects, particularly in comparison with its direct competitors on a global level (Japan and the US).The basic idea behind ERA was to provide favorable conditions for all steps of the innovation process, including research at RIs that seemingly possess a strong strategic importance for the achievement of the goals set forth in the ERA strategy (Hallonsten 2020a). ESFRI got the role of a discussion forum and consultancy body to the European Commission, with the primary mandate of preparing and publishing a RI roadmap, which came out in a first edition in 2006 and was followed by updates in 2008, 2010, 2016 and 2018 (Bolliger and Griffiths, ch 5 in this volume). The most evident result of its policy advisory role to the European Commission has been the launch of a completely new organizational form, ERIC, created in 2009 by European legislation (Moskovko et al 2019; Moskovko, ch 6 in this volume). Nonetheless, in spite of the emerging and growing policy role of the EU in collaborative large-scale research projects and the introduction of the ERIC legislation, these projects are still built on intergovernmental ad hoc agreements and voluntary participation (Krige 2003). The creation of the European X-Ray Free-Electron Laser (European XFEL) in 2009, the Facility for Antiproton and Ion Research (FAIR) in 2010 and the European Spallation Source (ESS) in 2009 are only a few of the most recent examples in this regard (Cramer, ch 3 in this volume; Hallonsten 2002b).

Reconciling from above, all these historical and recent examples of collaborative large-scale research in Europe have in common that alliance building has been key to bring the projects into being. Most of the projects were initiated by a very small number of countries, often the politically most powerful, such as France, (West) Germany or the United Kingdom, who entered into basic agreements to which other countries were invited to join. In the most recent years, large-scale research collaborations in Europe also increasingly include Central Eastern European countries formerly under communist rule, and countries from outside the EU or continental Europe, such as Russia, India and South Africa. The political realities of these collaborations often draw upon and reflect the (geo)political dynamics of Europe. It has been noted in several previous works that the tension between national interest and common good permeates the setting up and running of these scientific collaborations (Krige 2003; Hallonsten 2012, 2014; Cramer 2017). However, these considerations do not show any unequivocal trends: they are specific and differ according to the characteristics of each collaboration and negotiation process, such as the scientific purpose, political circumstances, historical context and technical challenges (Hallonsten 2012).

3. THE CONCEPT BIG SCIENCE

In a scholarly context, Big Science was first used in the 1960s, with two different meanings. First, Weinberg (1961, 1967), then director of the US-American Oak Ridge National Laboratory, used the concept Big Science (capitalized) to describe what he saw as a worrying development in the organization of R&D, connected to the Military-Industrial Complex, namely "an increasing subordination of individual scientists under large and bureaucratic projects, with a consequential damage to academic freedom and individual creativity" (Weinberg 1967: 9). Weinberg's argument was based on first-hand experiences from his own organization, the Oak Ridge National Lab, where weapons development and the operation of large infrastructures of modern physics (reactors and particle accelerators) dominated the scene. Second, physicist and historian of science Derek De Solla Price (1963/86) used big science (non-capitalized) to describe a seemingly exponential growth of science in the industrialized world, counted in funding, manpower and output.

Since then, the term and concept Big Science has evolved and become associated with all kinds of things. As noted by Capshew and Rader (1992: 18–19), only very few of the many studies that use "Big Science" in scholarly work relate to the original ideas of Weinberg or Price. Instead, Big Science has become conceptually very flexible and broad, and joined the group of similar popularized uses of the prefix "Big", such as *Big Business* (Fay 1912; Drucker 1947), *Big Government* (Pusey 1945), *Big Democracy* (Appleby 1945), *Big Cities* (Rogers 1971), *Big Foundations* (Nielsen 1972) and, most recently, *Big Pharma* (Law 2006; Ansell 2013) and *Big Data* (Cukier and Mayer-Schonberger 2013). All of these illustrate phenomena that are claimed to somehow have become *big* in their own way, in other words, it is postulated, by the use of the prefix, that the size and scale of these things are dramatically different than their prior versions. The fate of Big Science is similar to many of those other examples, namely a conceptual dilution. A broad community of scholars has squeezed all kinds of things into the concept Big Science: the space programs of the 1980s (Smith 1989; Kay 1994), the large corporate R&D divisions of the 1970s (Hounshell 1992), state-controlled research in the Soviet Union in the early twentieth century (Graham 1992; Kojevnikov 2002), large projects in biology, ecology and geoscience (Aronova et al 2010; Kevles 1997), nineteenth-century naturalist explorer missions to Latin America (Knight 1977) and sixteenth-century astronomy (Christianson 2000), besides the usual suspects of the wartime atomic bomb project in the US (the Manhattan Project) (Hughes 2002), telescopes for ground-based astronomy (McCray 2006) and accelerator complexes for particle physics (Galison et al 1992; Hiltzik 2015) and materials science (Westfall 2008; Hallonsten 2016a).

Catherine Westfall has adopted a more narrow perspective, arguing that the standard notion of "Big Science began in the radar and atomic bomb projects in World War II, prospered during the Cold War, and showed signs of decline at the end of that era, as signaled by the 1993 cancellation of the multibillion-dollar high-energy physics project the Superconducting Supercollider (SSC)" (Westfall 2003: 33). This view, however, has been challenged by several scholars, including Westfall herself, who have noted that Big Science in an institutional sense did not vanish with the end of the Cold War but rather adapted and took new shapes, to meet new expectations and demands in a post-Cold War era dominated not by nuclear research but advanced materials sciences, molecular biology and life sciences (Hallonsten and Heinze 2012, 2016; Westfall 2008; Jacob and Hallonsten 2012; Hallonsten 2016a; Crease and Westfall 2017). Elzinga concurs, arguing that while national defense was the primary motivation for Big Science in the Cold War era, the recent decades have been characterized by internationalization of science, including Big Science (Elzinga 2012).

Several authors have made attempts to conceptually define Big Science and thus contribute to a detailed and more nuanced understanding of it. The effort of Galison and Hevly (1992), gathering a handful of scholars in an edited book entitled *Big Science: The Growth of Large-Scale Research*, led to an expanded catalogue of examples but little progress in terms of conceptually framing the available theoretical and empirical work on Big Science as a phenomenon of recent and contemporary science and society. Capshew and Rader (1992) made an exhaustive inventory of the uses of Big Science in scholarly work, and thus produced a conceptual taxonomy with six main areas of the use of the term: (1) a pathological or (2) natural state of science, produced by historical inevitability (cf. Weinberg 1961, 1967; Price 1963/86); the conduct of science (3) with the use of especially big instruments, (4) in especially big ("industrial") organizational arrangements, or (5) in an institutionalized mode tied to contemporary society's need for symbols of progress; and (6) especially politically entangled science. Although their article is a key contribution to the historical contextualization and categorization of different forms of Big Science, it falls short of contributing to a nuanced conceptual framework or theoretical model that could guide contemporary studies and analyses of Big Science outside the history of science.

A recent sociological attempt to remedy this theoretical deficit produced the distinction between a "wide" and "narrow" interpretation of Big Science, where the former includes almost anything that has a connection to science and technology, and is big in one way or the other, and the latter defined in an analysis of some specific contemporary forms of Big Science (Hallonsten 2016a). Temporarily subscribing to this distinction, and dwelling for a moment in the analytical content of the "narrow" interpretation, can yield some insights concerning the typical problems of the concept of Big Science and how they can be handled. The "narrow" interpretation of Big Science corresponds to "science made big in three dimensions: big organizations, big machines, and big politics" (Hallonsten 2016a: 17).

Big machines refers to the size of scientific instruments, such as accelerators or telescopes, that often provide the (material) fundament of large-scale research projects. This view on what signifies Big Science is probably most intuitively evident for particle physicists and their increasing circumferences of circular-shaped particle accelerators, which were needed to reach ever higher energies and to explore and study ever smaller constituents of matter. However, alternative uses of accelerators for the production of x-rays and neutron beams that do not need such large circumferences have since long taken over at the expense of particle physics, which is key to the transformation of Big Science (in the "narrow" interpretation) that the definition was invented to assist in analyzing (Hallonsten 2016a: 5ff).

Big organizations means, for instance, the organization of large scientific projects in an industrial manner, or the hierarchical structure of large teams that were formed and organized around a large and costly scientific instrument which was (and still is) the case for e.g. particle physics or ground-based astronomy (Galison and Hevly 1992). This may also point to the increasing amount of financial and human resources devoted to facilitating experimental work, for example the organizations required to operate large instruments and infrastructure.

Yet, as already pointed out, an understanding of Big Science, which is mainly associated with the physical size of scientific instruments and machines and their supporting organizational frameworks, runs the risk of missing other kinds and forms of large-scale research. Moreover, it is difficult to define what counts as "big" compared to what has been large or big in the years and decades before. The US wartime nuclear bomb project, nicknamed the Manhattan Project, employed over 100 000 people at its height (Hughes 2002), whereas contemporary Big Science organizations such as the ESRF have permanent workforces of only a few hundred. Particle accelerators, the most ubiquitous examples of Big Science, vary greatly in size depending on areas of use and the historical era in which they were designed and built: The most advanced accelerator projects of the 1950s had circumferences of 50 to 100 meters; two decades later the cutting-edge facilities ran kilometer-long particle accelerators. Today, when the areas of use have shifted to x-rays and neutrons, circumferences and lengths are back at a couple of hundred meters (Hallonsten 2016a: 18–19). Nonetheless, "big machines" and "big organizations" as elements of a definition of Big Science also restricts the empirical scope, which can be useful for reasons of clarity and stringency: The purpose of the cited definitional work was to narrow down the analysis to those Big

Science endeavors that are physically bound to single infrastructural sites. In order to do useful empirical studies in other distinctly demarcated parts of the very broad spectrum of scientific projects, programs, infrastructures and instrumentation covered by the "wide" interpretation, quite obviously a corresponding degree of clarity and narrowness is necessary in the development of conceptual frameworks to be used for those purposes, albeit of course with the help of other criteria.

Big politics refers to the observation that large-scale research projects require quite substantial political intervention for their realization. Apart from the fact that any investment in scientific research without immediate practical benefit requires investment from the public purse, and thus enters a game of political priority setting, science that is perceived as "Big" is also visible, and positions science on the "maps of politicians and policymakers" (Hackett et al 2004: 748). This also leads to *expectations* on the political level, not only in relation to national security and geopolitical interest spheres (as during much of the Cold War), but also since science and technology have become increasingly associated with economic growth and competitiveness through the much praised issue of *innovation* (Berman 2012; Elzinga 2012). In other words, rhetoric to justify the continued investment and political support of Big Science has also been adapted to varying political agendas (Jacob and Hallonsten 2012: 412). After World War II, the beliefs and fears in the powers of nuclear physics seem to have been enough to propel Big Science into its own league of superpower competition, especially after the 1957 Sputnik crisis which extended the expense account of the federal US government significantly, also taking Europe with it in a vast expansion of science budgets across the board, and especially large-scale programs. In the 1960s and 1970s, when social movements protested against the power structures of the political and military establishments, large-scale governmental R&D and Big Science got its share of criticism. Towards the end of the Cold War, when the US invested heavily in the Strategic Defense Initiative, the SSC and several other projects, Big Science took a place at the centre of politics again (Smith 1990; Gaddis 1982/2005). In the post-Cold War era, when investments in Big Science are seemingly made on the basis of an expectation that they will contribute to gained or increased scientific/technological competitiveness, the political stakes are likewise high (Hallonsten 2016a: 185ff).

The politics of Big Science in Europe is differentiated just like any other area of politics on the continent; there is the national level and the intergovernmental/collaborative level. As discussed above, the traditional approach of European governments has been to regard collaboration and the joint funding and contribution to Big Science projects as a necessity in order to keep a competitive position in relation to the US and Japan, in areas such as astronomy, particle physics and the use of synchrotron radiation and neutrons for studies of materials. But larger European countries have also maintained national programs in these areas and continue to operate national Big Science facilities. It is also important to remember, when discussing the big politics of Big Science, that the funding for these projects is miniscule compared to the vast systems of publicly funded R&D in universities and institutes, which employ tens of thousands of people and share millions of euros every year also in smaller countries. Clearly, there are nuances and a depth in the meaning of "Big Science" that warrants further scholarly analysis, theoretically and empirically.

4. THE CONCEPT RESEARCH INFRASTRUCTURES

While there is abundance of the use of Big Science in scholarly works, Research Infrastructures is not a widely used term among historians and sociologists. The first scholarly use of the term seems to be by Papon (2004), and the first comprehensive analysis of the concept, its origins and its use suggests that the origins and motivation for the term root in political need or ambition to (re)brand certain projects and investments to distinguish them from other investments in science and to highlight their boundary-spanning and facilitating nature (Hallonsten 2020a). Meanwhile, contrary to Big Science, there are definitions of RIs available, some of which are also sanctioned by various political organs and governance bodies.

However, concerning definitions, it is important to separate the current use of the term Research Infrastructures by policymakers of the EU and several national governments, and its growing use in scholarly works (see below), from a more general use of "research infrastructure" or plainly "infrastructure" for research, in other instances or further back in time. As already noted, analogous to the suggestion of Capshew and Rader (1992) to separate "Big Science" from "big science", it is possible to distinguish between "Research Infrastructures" as a specific policy concept with wide use especially in the EU's current research and innovation policy work, and "research infrastructures" as a broader term with a general or colloquial meaning, based on a commonplace understanding of "infrastructure" as resources (often physical, but sometimes organizational) necessary for some kind of operation or activity. The definition of "infrastructure" provided by the *Oxford Dictionary of English* is: "basic physical and organizational structures and facilities [...] needed for the operation of a society or enterprise". Adapting this definition, a general or colloquial meaning of "research infrastructure" could be basic physical and organizational structures and facilities needed for scientific work, which also probably corresponds to the occasional use of "research infrastructure" in a less specific and more colloquial meaning, for example, in Guzzetti's (1995) comprehensive history of European research policy, where the term is used once in connection with the general need to build up capacity in Europe after World War II, to keep up in the competition against the US.

But also with this general or colloquial definition, the term and concept "research infrastructure" is conspicuously absent from historical works that deal with basic physical and organizational structures and facilities needed for science, such as particle accelerators, astronomical observatories, and other larger equipment. The history of science is filled with accounts on, and analyses of, the planning, funding, construction, operation and use of such "research infrastructures" without ever using this term (see e.g. Cathcart 2006; Strasser 2011; Nye 1996; Livingstone 2003; Agar 2012; Kevles 1977/95; Pais 1986; Martin 2018; Mody 2011). The key works that identify, classify and define the use of large equipment and instrumentation in science do not use the term at all (e.g. Capshew and Rader 1992; Galison and Hevly 1992; Westfall 2003). Thus we can conclude that in the general meaning, and without capitalization, "research infrastructures" is really not a distinct term and concept at all. The reason is probably that any analytical effort that attempts to build on a general definition of "research infrastructure", for instance the one obtained by the adaptation of the definition of "infrastructure" from the *Oxford Dictionary of English* (above), yields a concept and term which is unworkably vague: anything can be a research infrastructure and a research infrastructure can be anything. Indeed, the laptop used to write this text, the funding body that provided support for this book project and the collection of books and articles cited in this chapter and throughout this book are all physical or organizational "research infrastructures" in a wide meaning.

Moving on to "Research Infrastructures" as a concept of policy origin, there is much to suggest that the *raison d'être* of this concept lies (so far) in political convenience, opportunism or necessity, rather than the identification of a clearly marked organizational field or sector of any national or international science and innovation system. Until recently, the concept "Research Infrastructures" has been used solely in a European context. RIs have been declared a "pillar" of the ERA strategy by the European Commission and "the key drivers for European capacity building" (European Commission 2008: 9, 15). ESFRI, which was established in 2002 "to support a coherent and strategy-led approach to policy making on [RIs] in Europe" (ESFRI 2016: 4), highlights the strategic importance of RIs for science and innovation in Europe, claiming that they "provide unique opportunities for world-class research and training" (ESFRI 2006: 10). ESFRI also stresses that RIs are "vital to make the European Research Area attractive at a global level" (ESFRI 2008: 5), and that they are "a key instrument in attracting and bringing together researchers, funding agencies, politicians and industry to act together and tackle the cross-disciplinary scientific and technical issues of critical importance for our continued prosperity and quality of life" (ESFRI 2010: 7).

ESFRI's definition of RIs is "facilities, resources or services of a unique nature that have been identified by European research communities to conduct top-level activities in all fields" (ESFRI 2010: 7). This is clearly a *political* rather than *analytical* definition: Few, if any, of the projects listed in the ESFRI roadmaps are "unique" in the original sense of the word, but have direct counterparts and competitors in the US and Japan, and also across Europe. These counterparts and competitors may, in certain aspects, outperform the ESFRI-listed projects, which also means that "top level" is imprecise and normative or politically motivated rather than descriptive. Moreover, while the ESFRI roadmaps and the lists of projects included in them have been drafted in peer-review committees consisting of top representatives of scientific communities, it is not clear either from the roadmaps or from the definition itself whether any political prioritization process has also been involved. In other words, it is unclear if the projects listed in the roadmaps have been selected solely on the basis of a stringent evaluation of performance or potential according to specific criteria or functions, sizes or reaches in use; what exactly these criteria are; or how they have been established (Bolliger and Griffiths, ch 5 in this volume).

Clearly, the EU places great trust, at least rhetorically, in the capacity of RIs to be key loci of cross-border collaboration in European science and innovation, and key drivers of an enhancement of research and innovation in service of the common market and to solve current and future challenges to sustainability, health and quality of life. Similar statements can be found in the several roadmap documents and reports developed and published by governments and research funding agencies in a handful of European countries, including Germany (BMBF 2013), Sweden (Swedish Research Council 2008) and Denmark (Strategiske Forskningsråd 2005) (see Bolliger and Griffiths, ch 5 in this volume).

The popularity of the term and concept Research Infrastructures in this EU-related context links to the increasing importance that science policy documents, such as reports from ESFRI and the Organisation for Economic Co-operation and Development's (OECD) Global Science Forum, attribute to the role of research infrastructures (non-capitalized) for economic competitiveness. In 2000, the EU launched its "Lisbon Strategy" with new directions to the economic policy of the EU and the strategic goal that Europe should "become the most competitive and dynamic knowledge-based economy in the world, capable of sustainable economic growth with more and better jobs and greater social cohesion" by 2010. The Lisbon strategy endorsed the ERA initiative that had been launched two months before and that can be interpreted as a "third generation" of EU policy involvement in science and innovation, corresponding to a transition from science policy (first generation) to technology policy (second generation), and finally to innovation policy (Borrás 2003). The approach to RIs in the ERA framework is based on the objective to consolidate common resources to conduct research, not only in natural sciences, but also in other fields such as the humanities and social sciences. One goal of ERA is to "develop world-class research infrastructures" and to "optimise the use and development of the best research infrastructures existing in Europe" (European Commission 2008). The several initiatives in the years thereafter, including the creation of ESFRI and its publication of several consecutive roadmaps, the funding programs for RIs within the Framework Programmes and the creation of the ERIC organizational form by EU regulation in 2009, should be seen in light of the very ambitious goals for RIs laid down in the ERA policies (see above and Ulnicane, ch 4 in this volume; Bolliger and Griffiths, ch 5 in this volume; Moskovko, ch 6 in this volume).

Approaching the same issue from another angle, there is much to suggest that the political hype in Europe and the EU around RIs in the past two decades is largely due to new patterns of research funding and organization. With the increased competition for funds and the "economization" (Berman 2014) and "commodification" (Radder 2010) of academic science, block-grant funding to universities and other research environments has been replaced with line-item funding with strings attached and tighter control mechanisms for resource utilization. Among other things, this has meant the downsizing of many of the resources built, maintained and operated internally at university faculties, departments and institutes, such as larger laboratory equipment, instrumentation for analysis, databases and sample repositories, but also (comparably) smaller particle accelerator facilities and similar. In order to secure funding for these technical assets, that are absolutely crucial for much scientific work, and in order to secure sustainable organizational arrangements for their maintenance, development and use, specific funding programs and specific policy instruments needed to be put in place. From the side of the projects and practitioners, it can be noted that several of the RIs on the ESFRI roadmap are of a size and scope that typically would have made them into parts of academic environments or institutes some decades ago, which the science funding systems of the time also enabled.

A prime example is the European Social Survey, initiated by the European Science Foundation in 1996 to enable the pan-European biannual collection and handling of data on social issues, and operated by a network of universities and institutes in Europe. The European Social Survey was struggling with organizational framework and the lack of a stable and sustainable funding solution until it was included in the 2006 ESFRI roadmap with an estimated "construction cost" of €9 million to cover for the setting up of a more robust organization and funding structure. In 2013, the European Social Survey became an ERIC (Duclos Lindstrøm and Kropp 2017). It is quite clear that the project's identification as a "research infrastructure" in the social sciences, and its *consecration* to the status of a European research infrastructure (see below), in other words of "unique nature" and "identified by European research communities to conduct top-level activities" (ESFRI 2010: 7), was used by its managers and champions to rescue the project and secure a longer-term and more stable funding structure (Duclos Lindstrøm and Kropp 2017: 861).

The variety within the collection of 60 facilities, databases, networks, vessels, etc. that have been listed on any of the four editions of the ESFRI roadmap to date, and/or that have been granted ERIC status by the European Commission, is huge: We find here the ESS and European XFEL, both single-sited, accelerator-based facilities (and prime examples of what has been called "transformed" Big Science, see above) with construction costs close to €2 billion; the European Social Survey, which is a data collection exercise and repository of significantly smaller size and no physical infrastructure; the Integrated Carbon Observation System and the European Multidisciplinary Seafloor and Water Column Observatory which are collections of utilities for the observation of various patterns of environmental change and the Central European Research Infrastructure Consortium which is only a network of pre-existing research facilities that coordinates experimental work and provides users with a single entry point. As shown in further detail by Hallonsten (2020a) and Moskovko (ch 6 in this volume), these projects and organizations differ greatly in several respects: size, cost, organizational form, user communities, scientific areas served and the breadth and size of these.

The question remains what the criteria are for identifying projects and organizations as RIs, and who decides. The costliest of all RIs listed in an ESFRI roadmap document and/or granted ERIC status is the ESS, under construction in southern Sweden. It is certainly, by most relevant standards, possible to identify as Big Science, and nowadays it also has "Research Infrastructure" in its official name, since it is an ERIC, but it did not self-identify as a "research infrastructure" but rather "research facility" in its annual reports up until the application for ERIC status in 2015. Nonetheless, it has been listed in all the ESFRI roadmaps and in all editions of the Swedish national roadmap (e.g. Swedish Research Council 2008). Meanwhile, there are of course several resources of various types and in many different organizational settings that would qualify as "research infrastructures" simply by being identical or very similar to the projects on the ESFRI roadmap (or any national roadmap of a European country) and by fulfilling the very general definitions cited above, but that do not self-identify as such and/or have not been branded as such by ESFRI or the European Commission or any national policymaking body.

Contrary to Big Science – which is a vague and broad term and concept because scholars and policymakers have been in the habit of using it for whatever purposes they want – Research Infrastructures in the narrower, political sense (and capitalized) therefore appears as binary: Either you're self-identifying, or branded by someone else, as RI, or you are not. This binarity seems largely detached from any analytically useful categorizations of size, cost, organizational forms, use, communities and so on – also here, anything can be a Research Infrastructure and a Research Infrastructure can be anything, provided that the label is affixed to this *anything*. ESFRI, the European Commission and several national bodies (see Bolliger and Griffiths, ch 5 in this volume) have the ability to *consecrate* almost anything to a status of RI. The origin of the term *consecration* is in the Roman Catholic Church where it literally means "association with the sacred", in other words the rite of making something holy. It was first used in a sociological context by Bourdieu (1975, 1988) to denote the inclusion of individuals in scientific elites by the awarding of several forms of social recognition, and more recently by Holmqvist (2017) in a study of how people become part of social elites in Sweden. The reader is well served by bearing in mind, when proceeding through the collection of chapters that makes up this book, that such consecration of a project does not mean either membership in any specific association or field of entities or organizations, or that the project automatically fulfils some predefined criteria. "RI" is a label, just like "Big Science" and it can be used on almost anything, provided that there is a will to do so.

5. CONCEPTUAL LANDSCAPE

An important question has been lurking in the background of the previous sections, namely how the concepts Big Science and RIs relate to each other. The issue is not easily resolved; the analytical vagueness and availability of different interpretations depending on viewpoint and ambitions makes the task of positioning the two in some analytical relationship thorny. In a most rudimentary sense (and keeping the concepts non-capitalized), it can be concluded that some research infrastructures are big science, whereas some are not, at least not relative to others. It can also be noted that the term Research Infrastructures (capitalized) has so far been used mostly in a European context, which means that based on the argument about binarity and *consecration* at the end of the previous section, many US-American facilities and projects should not be branded as RIs even if their direct European counterparts are. In principle, the same is true also for any other example where the term research infrastructure (non-capitalized) or Research Infrastructure (capitalized) is not used, such as in those countries where roadmaps identify "large facilities" or "large equipment" or similar but, for any reason, avoid calling them "research infrastructures".

Big Science and RIs most evidently overlap where large facilities such as particle accelerators are used for scientific research by a community that extends beyond the inhouse scientists of the organization that operates the facilities: ESRF and the European XFEL are prime examples (Cramer 2020). Second, there are cases of Big Science that are (probably) not RIs. The Human Genome Project, undertaken in 1990–2003, has been called Big Science and compared to the SSC by Kevles (1997), but was certainly not an RI (although it made use of vast instrumentation) but rather a *research project*. Third, there are cases of RIs that are not big science which is more difficult to define since "big" in the wide interpretation of "big science" is relative and a very inclusive concept. But it may well be argued that the European Social Survey (see previous section) is not at all big science, to say nothing of the many smaller projects that national RI roadmaps frequently highlight and that are minuscule in size and cost.

Other concepts have been in sway in the communities that have studied big science and research infrastructures, most notably "megascience" which has at least two meanings. First, there was the OECD Megascience Forum which was a discussion forum like ESFRI, organized by the OECD that met regularly between 1992 and 1999 to foster cooperation among OECD countries in the planning, construction and operation of large research facilities. It is not clear why the forum used the term "megascience" and not simply "big science", but it is probably not a far-fetched guess that the reasons were rhetorical and political. There were also no clear definitions, and so "megascience" in this context was, like "Research Infrastructures", a politically motivated term. Second, "megascience" has also been used with a sociological/analytical meaning. The history of Fermilab, the US-American flagship particle physics facility in Illinois, inaugurated in 1972, was chronicled and analyzed by Hoddeson et al (2008) who identified the transition of particle physics in the 1970s into a new state as a move from "big science" to "megascience". The development had a micro and a macro component, interrelated of course. On the micro level, the experiments in particle physics grew complex, and stretched out in time, to the degree that instrumentation became too big and costly for ordinary research groups to fund and operate, which led to the growth of teams to hundreds (and later thousands) of researchers and engineers, and the prolongation of experiments for several years. The macro-level development came from the need to concentrate resources to fewer facilities and lab organizations, given the slowdown of economic growth in the early 1970s and the dramatic increase in the size and cost of particle physics facilities and instruments. The "megascience" development has continued; if the gargantuan SSC would not have been cancelled by the US Congress in 1993, it would certainly have monopolized the US federal particle physics budget and produced the closedown of similar activities at Fermilab and elsewhere. In the early 2000s CERN completed its construction of the Large Hadron Collider, with involvement by US-American, Japanese, Chinese and Russian particle physicists, and became the only particle physics laboratory in the world that operates a major experimental facility.

The above described micro-level "megascience" trend is also manifested at CERN; the two publications that announced the discovery of the Higgs boson in 2012 (with the use of the Large Hadron Collider) had close to 3000 authors each – experiment teams in "megascience" are nowadays counted not by the hundreds, but by the thousands. But this is a phenomenon isolated to particle physics – no other science has had a similar development, and other forms of Big Science are still serving teams of researchers by the size of five to ten, or occasionally, 40–50 (Hallonsten 2016a: 100). It can, hence, be concluded that "megascience", at least in the analytical meaning of the word as developed by Hoddeson et al (2008) and further refined by Hallonsten (2016a: 62–4), is a particle physics phenomenon.

A related concept is "megaprojects", which has gotten its own subspecialism of macrosociology, much by the work of Flyvbjerg (2003, 2008, 2014, 2017). The literature on megaprojects is rich in detail and analytical ambitions, but never deals specifically with megaprojects in science, although at least some of the projects typically identified as Big Science, and some of the largest projects consecrated to RI status by ESFRI and the European Commission, certainly fit the definition of megaproject as provided by Flyvbjerg (2017: 2): "large-scale, complex ventures that typically cost $1 billion or more, take many years to develop and build, involve multiple public and private stakeholders, are transformational, and impact millions of people". Save for the final part – Big Science projects do not impact millions of people other than in an indirect sense, by their general contributions to scientific progress and thus, by extension, to society – this definition applies to the typically mentioned "classics" of Big Science in Europe, if the $1 billion threshold is adjusted for inflation: CERN, ESO, ILL, ESRF, XFEL and ESS. Importantly, however, it does not apply to most RIs, which is yet another reminder of the enormous variety within the field of projects, organizations and entities that this book is about. The chapter by Hallonsten (ch 10 in this volume) explores the (dis)similarities between "megaprojects" and Big Science or RIs further, by analyzing if Flyvbjerg's "iron law" of megaprojects – that they always exceed budgets and timetables – holds for Big Science.

It is also necessary to mention some of the juxtapositions of the concepts – a major theme in the analysis that produced the conceptualization of "transformed Big Science" (Hallonsten 2016a) was the identification that a lot of contemporary big science is really "small science on big machines", i.e. that contrary to the developments in particle physics that has been conceptualized as "megascience" (above), the currently most extensive use of particle accelerators is for the production of x-rays and neutrons which are experimental resources used by ordinary research groups from universities, institutes and, occasionally, industrial firms, which are of quite ordinary sizes and only use x-rays and neutrons (and thus the instrumentation at the Big Science facilities to which they travel for shorter visits) as resources among several others, not least including laboratory equipment in their home organizations. It deserves to be noted that the disciplinary breadth of these user communities is wide and growing, and that especially applications in biology and the life sciences have grown tremendously since the early 1990s (Hallonsten 2016b).

Although this conceptualization of a "transformed Big Science" builds on the "narrow" interpretation (see above) and thus restricts itself to the use of particle accelerators and reactors for experimental science, there is certainly a connection to be made to RIs. One of the things that many of the entities and projects consecrated to the status of RIs by the EU or national actors have in common is that they are resources for scientific research used by quite ordinary research groups in universities, institutes and industrial firms. This, furthermore, seems to fit quite well with the policy objectives of ESFRI, namely to let RIs "provide unique opportunities for world-class research and training" (ESFRI 2006: 10). This should also be borne in mind in the following – part of the continued success of Big Science beyond the Cold War context, and the rise of RIs as a relevant policy concept, lies in the broadening of the uses of Big Science to new scientific fields with clearer and more direct practical applications.

6. THE BOOK

The collection of chapters that make up this book have been written by scholars across Europe who share an interest in the topics of Big Science and RIs as outlined in the previous sections, and a determination to advance and improve our understanding of these issues by adding more stringent scholarly analysis.

Nicolas Rüffin's effort, in Chapter 2, is a model for everyone with such ambitions. In his well-needed and unprecedented review of methods and approaches in the very amorphous field of study that this book writes itself into, he provides a comprehensive literature list which in itself is a formidable contribution, but also an analysis of the field as such, based on approaches thus far tried. Important and interesting to note is also the reconciliatory effort made by Rüffin in this chapter concerning the incoherencies and contradictions in studies on Big Science and RIs in terms of theoretical and empirical scope and focus. The pragmatic solution to the complicating variety of cases and categories of Big Science and RIs that he offers when pointing to the need of openness and flexibility in order to learn and gain deeper and broader knowledge is instructive.

Chapter 3, by Katharina Cramer, deepens and widens the historical context of Big Science and RIs by demonstrating how the pursuit of scientific and technological excellence in Europe in the twentieth century has taken place within the framework of a broader political development of integration lined with crises and successes. The political integration process of Europe since World War II has been mirrored in the history of collaborative Big Science in Europe and the several projects highlighted by Cramer in her chapter have been very much shaped by their times, but also sometimes functioned as political and diplomatic spearheads for political developments with far broader importance and meaning.

In Chapters 4, 5, and 6, the role of the EU in the two most recent decades of research and innovation policy generally, and RI policy specifically, is analyzed in depth. Chapter 4, by Inga Ulnicane, discusses both the overall level of top EU initiatives and ambitions with a general ambition, and how these have played out on the level of a specific case, the Human Brain Project. Chapter 5, by Isabel Bolliger and Alexandra Griffiths, chronicles the genesis and development of ESFRI and its roadmap, and also provides a survey of the landscape of RI roadmapping processes in national contexts in Europe. In Chapter 6, Maria Moskovko traces the origins of the ERIC regulation and the entirely new ERIC organizational form, and also discusses how it has acted out in practice, in a comprehensive survey of the 21 ERICs that exist to date.

Chapter 7, by Thomas Franssen, is a much-needed case study of RIs in the humanities. Franssen shows, with the help of a competent analysis of policymaking in the Netherlands in the past few decades, how the concept of RIs has been used as a policy tool, to revitalize the area of humanities by steering its focus towards tangible (and thus measurable) activities that, from a policymaker's point of view, appear as more useful or relevant. The chapter fulfills the dual purpose of giving a crucial insight into how RI policy functions in the area of humanities, and how RI funding is used as a policy instrument.

Chapters 8–11 each deal with different aspects of Big Science and RIs that are all highly relevant and explanatory in the context of this book. Chapter 8, by Olof Hallonsten, Hjalmar Eriksson, and August Collsiöö, argues that RIs should not be handled in policy or funding contexts as scientific production units on par with university departments or institutes, since their role is essentially different. With the use of the classic sociological concept of functional differentiation, and an in-depth case study of a research infrastructure (non-capitalized) that has been around since the 1950s, the authors of this chapter make the case that a systems view on innovation enables the identification of RIs as a distinct type of resource or unit that needs to be handled as such in policymaking. Chapter 9, by Andrew Williams and Jean-Christophe Mauduit, takes a comprehensive grip on the issue of access for researchers to RIs and how this relates to national and international policy considerations and, crucially, the issue of socio-economic impact and usefulness of Big Science and RIs. As they show, the issue of access is not only absolutely vital, but indeed the basic *raison d'être* for the existence of these RIs, and therefore a topic in great need of analysis. Chapter 10, by Olof Hallonsten, applies the concept of "megaprojects" and the "iron law" of cost and timetable overruns of "megaprojects" on Big Science and RIs, using this as the basis for a discussion on the challenges built into the very heterogeneous planning processes and long time horizons that characterize Big Science projects. Similarly taking a longer time perspective, Chapter 11, by Beatrice D'Ippolito and Charles-Clemens Rüling, shows how a Big Science facility can manage to renew itself and maintain a central position in an international (European) scientific community, also contrasting this role as stable and enduring center of gravity for a dynamic and changeable user community with technical limitations and what this means for the life cycle of the RI. The chapter also demonstrates how an RI can remain the center of, and to some extent engine of, a broad and inclusive European user community, thus to some extent also lending substance to the claim that RIs can be "the key drivers for European capacity building" (European Commission 2008: 15).

Such substance to otherwise comparably hollow claims is one of the main aims of this book. The concluding chapter will reflect on the fulfillment of this aim and summarize the main findings, so that the reader is oriented in the state of the art of the study of Big Science and RIs in Europe, established or updated by the collected chapters in the book, to the benefit of scholars, policymakers and the interested public.

REFERENCES

Agar J (2012) *Science in the twentieth century and beyond.* Polity Press.

Ansell J (2013) *Transforming big pharma.* Gower Publishing.

Appleby P H (1945) *Big democracy.* Knopf.

Aronova E, K S Baker and N Oreskes (2010) Big science and big data in biology: From the international geophysical year through the international biological program to the Long Term Ecological Research (LTER) network, 1957–present. *Historical Studies in the Natural Sciences* **40** (2): 183–224.

Berman E P (2012) *Creating the market university: How academic science became an economic engine.* Princeton University Press.

Berman E P (2014) Not just neoliberalism: Economization in US science and technology policy. *Science, Technology, and Human Values* **39** (3): 397–431.

Blaauw A (1991) *ESO's early history.* ESO.

BMBF (2013) Roadmap for research infrastructures: A pilot project of the Federal Ministry of Education and Research (BMBF). Available at https://ec.europa.eu/research/infrastructures/pdf/roadmaps/germany_national_roadmap_en.pdf (last accessed December 15, 2019).

Borrás S (2003) *The innovation policy of the European Union: From government to governance.* Edward Elgar Publishing.

Bourdieu P (1975) The specificity of the scientific field and the social conditions of the progress of reason. *Social Science Information* **14** (6): 19–47.

Bourdieu P (1988) *Homo academicus.* Stanford University Press.

Capshew J H and K A Rader (1992) Big science: Price to the present. *Osiris* 2nd series, **7**: 3–25.
Cathcart B (2006) *The fly in the cathedral*. Farrar Straus and Giroux.
Christianson J R (2000) *On Tycho's island: Tycho Brahe, science, and culture in the sixteenth century*. Cambridge University Press.
Cramer K C (2017) Lightening Europe: Establishing the European Synchrotron Radiation Facility (ESRF). *History and Technology* **33** (4): 396–427.
Cramer K C (2020) *A political history of Big Science: The other Europe*. Palgrave Macmillan.
Crease R and C Westfall (2017) The new Big Science. *Physics Today* **69** (5): 30–6.
Cukier K and V Mayer-Schonberger (2013) *Big data: A revolution that will transform how we live, work and think*. Murray.
D'Ippolito B and C-C Rüling (2019) Research collaboration in Large Scale Research Infrastructures: Collaboration types and policy implications. *Research Policy* **48**: 1282–96.
Drucker P F (1947) *Big business*. Heinemann.
Duclos Lindstrøm M D and K Kropp (2017) Understanding the infrastructure of European Research Infrastructures: The case of the European Social Survey (ESS-ERIC). *Science and Public Policy* **44** (6): 855–64.
Elera Á De (2006) The European Research Area: On the way towards a European scientific community? *European Law* **12** (5): 559–74.
Elzinga A (2012) Features of the current science policy regime. *Science and Public Policy* **39** (4): 416–28.
Elzinga A and A Jamison (1995) Changing policy agendas in science and technology. In S Jasanoff, G E Markle, J C Petersen and T Pinch (eds) *Handbook of science and technology studies*. Sage, pp 572–97.
ESFRI (2006) Strategy report on research infrastructures. Roadmap 2006.
ESFRI (2008) Strategy report on research infrastructures. Roadmap 2008.
ESFRI (2010) Strategy report on research infrastructures. Roadmap 2010.
ESFRI (2016) Strategy report on research infrastructures. Roadmap 2016.
European Commission (2008) *Developing world-class research infrastructures for the European Research Area (ERA): Report of the ERA expert group*. Office for Official Publications of the European Communities.
European Council (2009) Council Regulation No 723/2009 of 25 June 2009 on the Community legal framework for a European Research Infrastructure Consortium (ERIC) [2009] OJ L 206/1, as later amended by Council Regulation (EC) No 1261/2013 of 2 December 2013 [2013] OJ L 326/1.
Fay C N (1912) *Big business and government*. Moffat, Yard and Co.
Flyvbjerg B (2003) *Megaprojects and risk: An anatomy of ambition*. Cambridge University Press.
Flyvbjerg B (2008) *Decision-making on mega-projects: Cost-benefit analysis, planning and innovation*. Edward Elgar Publishing.
Flyvbjerg B (2014) *Megaproject planning and management: Essential readings*. Edward Elgar Publishing.
Flyvbjerg B (2017) *The Oxford handbook of megaproject management*. Oxford University Press.
Gaddis J L (1982/2005) Strategies of containment: A critical appraisal of American national security policy during the Cold War, 2nd ed. Oxford University Press.
Galison P and B Hevly (eds) (1992) *Big science: The growth of large-scale research*. Stanford University Press.

Galison P, B Hevly and R Lowen (1992) Controlling the monster: Stanford and the growth of physics research, 1935–1962. In P Galison and B Hevly (eds) *Big science: The growth of large-scale research*. Stanford University Press, pp 46–77.

Giroux H A (2007) *The university in chains: Confronting the military-industrial-academic complex*. Paradigm Publishers.

Graham L R (1992) Big science in the last years of the Soviet Union. *Osiris 2nd series* **7**: 49–71.

Greenberg D (1967/99) *The politics of pure science*, 2nd ed. University of Chicago Press.

Guston D H (2000) *Between politics and science: Assuring the integrity and productivity of research*. Cambridge University Press.

Guzzetti L (1995) *A brief history of European Union research policy*. European Communities.

Hackett E, D Conz, J Parker, J Bashford and S DeLay (2004) Tokamaks and turbulence: Research ensembles, policy and technoscientific work. *Research Policy* **33**: 747–67.

Hallonsten O (2012) Continuity and change in the politics of European scientific collaboration. *Journal of Contemporary European Research* **8** (3): 300–19.

Hallonsten O (2014) The politics of European collaboration in Big Science. In M Mayer, M Carpes and R Knoblich (eds) *The global politics of science and technology*, Vol. 2. Springer, pp 31–46.

Hallonsten O (2016a) *Big Science transformed: Science, politics and organization in Europe and the United States*. Palgrave Macmillan.

Hallonsten O (2016b) Use and productivity of contemporary, multidisciplinary Big Science. *Research Evaluation* **25** (4): 486–95.

Hallonsten O (2020a) Research Infrastructures in Europe: The hype and the field. *European Review* **28** (4): 617–35.

Hallonsten O (2020b) *The campaign: How a European Big Science facility ended up on the peripheral farmlands of southern Sweden*. Arkiv Academic Press.

Hallonsten O and T Heinze (2012) Institutional persistence through gradual adaptation: Analysis of national laboratories in the USA and Germany. *Science and Public Policy* **39**: 450–63.

Hallonsten O and T Heinze (2016) "Preservation of the laboratory is not a mission". Gradual organizational renewal in national laboratories in Germany and the United States. In T Heinze and R Münch (eds) *Innovation in science and organizational renewal: Historical and sociological perspectives*. Palgrave Macmillan, pp 117–45.

Heinze T, O Hallonsten and S Heinecke (2015a) From periphery to center: Synchrotron radiation at DESY, Part I: 1962–1977. *Historical Studies in the Natural Sciences* **45** (3): 447–92.

Heinze T, O Hallonsten and S Heinecke (2015b) From periphery to center: Synchrotron radiation at DESY, Part II: 1977–1993. *Historical Studies in the Natural Sciences* **45** (4): 513–48.

Heinze T, O Hallonsten and S Heinecke (2017) Turning the ship: The transformation of DESY, 1993–2009. *Physics in Perspective* **19**: 424–51.

Hermann A, J Krige, U Mersits and D Pestre (eds) (1987) *History of CERN. Volume I: Launching the European organization for nuclear research*. North-Holland.

Hermann A, J Krige, U Mersits and D Pestre (eds) (1990) *History of CERN. Volume II: Building and running the laboratory, 1954–1965*. North-Holland.

Hiltzik M (2015) *Big science: Ernest Lawrence and the invention that launched military-industrial complex*. Simon and Schuster.

Hoddeson L, A Kolb and C Westfall (2008) *Fermilab: Physics, the frontier, and megascience*. University of Chicago Press.
Holmqvist (2017) *Leader communities: The consecration of elites in Djursholm*. Columbia University Press.
Hounshell D A (1992) Du Pont and the management of large-scale research and development. In P Galison and B Hevly (eds) *Big Science: The growth of large-scale research*. Stanford University Press.
Hughes J (2002) *The Manhattan project: Big Science and the atom bomb*. Icon Books.
Jacob M and O Hallonsten (2012) The persistence of Big Science and megascience in research and innovation policy. *Science and Public Policy* **39**: 411–15.
Jacrot B (2006) *Des neutrons pour la science: Histoire de l'Institut Laue-Langevin, une coopération internationale particulièrement réussie*. EDP Sciences.
Kay W D (1994) Democracy and super technologies: The politics of the space shuttle and space station freedom. *Science, Technology, and Human Values* **19**: 131–51.
Kevles D J (1977/95) *The physicists: The history of a scientific community in modern America*. Harvard University Press.
Kevles D J (1997) Big Science and big politics in the United States: Reflections on the death of the SSC and the life of the Human Genome Project. *Historical Studies in the Physical Sciences* **27** (2): 269–97.
Knight D M (1977) *The nature of science: The history of science in Western culture since 1600*. Deutsch.
Kojevnikov A (2002) The Great War, the Russian Civil War, and the invention of big science. *Science in Context* **15** (2): 239–75.
Krige J (ed) (1996) *History of CERN*, Vol. III. North-Holland.
Krige J (2003) The politics of European scientific collaboration. In J Krige and D Pestre (eds) *Companion to science in the twentieth century*. Routledge, pp 897–918.
Krige J (2006) *American hegemony and the postwar reconstruction of science in Europe*. MIT Press.
Krige J and L Guzzetti (eds) (1997) *History of European scientific and technological cooperation*. European Communities.
Law J (2006) *Big pharma: How the world's biggest drug companies control illness*. Carroll and Graf.
Livingstone D N (2003) *Putting science in its place: Geographies of scientific knowledge*. University of Chicago Press.
Madsen C (2012) *The jewel on the mountaintop: The European Southern Observatory through fifty years*. Wiley.
Martin J (2018) *Solid state insurrection: How the science of substance made American physics matter*. Pittsburgh University Press.
McCray W P (2006) *Giant telescopes: Astronomical ambition and the promise of technology*. Harvard University Press.
Misa T and J Schot (2005) Inventing Europe: Technology and the hidden integration of Europe. *History and Technology* **21** (1): 1–19.
Mody C (2011) *Instrumental community: Probe microscopy and the path to nanotechnology*. MIT Press.
Moskovko M, A Astvaldsson and O Hallonsten (2019) Who is ERIC? The politics and jurisprudence of a governance tool for collaborative European research infrastructures. *Journal of Contemporary European Research* **15** (3): 249–68.
Nielsen W (1972) *The big foundations*. Columbia University Press.
Nye M J (1996) *Before Big Science: The pursuit of modern chemistry and physics, 1800–1940*. Harvard University Press.

Pais A (1986) *Inward bound: Of matter and forces in the physical world.* Oxford University Press.
Papon P (2004) European scientific cooperation and research infrastructures: Past tendencies and future prospects. *Minerva* **42** (1): 61–76.
Papon P (2009) Intergovernmental cooperation in the making of European research. In H Delanghe, U Muldur and L Soete (eds) *European science and technology policy: Towards integration or fragmentation?* Edward Elgar Publishing, pp 24–43.
Papon P (2012) L'Espace Européen de la Recherche (1960–1985): Entre science et politique. In C Defrance and U Pfeil (eds) *La construction d'un Espace Scientifique Commun? La France, la RFA et l'Europe après le "Choc du Spoutnik".* P I E Peter Lang, pp 37–54.
Pestre D (1997) Prehistory of the Franco-German Laue-Langevin Institute. In J Krige and L Guzzetti (eds) *History of European scientific and technological cooperation.* Office for Official Publications of the European Communities, pp 137–44.
Peterson J (1997) Eureka: A historical perspective. In J Krige and L Guzzetti (eds) *History of European scientific and technological cooperation.* Office for Official Publications of the European Communities, pp 323–45.
Price, D dS (1963/86) *Little science, big science ... and beyond.* Columbia University Press.
Pusey M J (1945) *Big government: Can we control it?* Harper & Bros.
Radder H (ed) (2010) *The commodification of academic research: Science and the modern universities.* Harvard University Press.
Rogers D (1971) *The management of big cities, interest groups, and social change strategies.* Sage.
Roland J-L (1997) COST: An unexpected successful cooperation. In J Krige and L Guzzetti (eds) *History of European scientific and technological cooperation.* Office for Official Publications of the European Communities, pp 355–68.
Simoulin V (2012) *Sociologie d'un grand équipement scientifique: Le premier synchrotron de troisième génération.* ENS Éditions.
Smith B (1990) *American science policy since World War II.* Brookings.
Smith R W (1989) *The space telescope: A study of NASA, science, technology, and politics.* Cambridge University Press.
Strasser B (2011) The experimenter's museum: GenBank, natural history, and the moral economies of biomedicine. *Isis* **102**: 60–96.
Strategiske Forskningsråd (2005) Fremtidens forskningsinfrastruktur – kortlægning af behov og forslag til strategi. Baggrundsrapport.
Swedish Research Council (2008) *Swedish Research Council's guide to the infrastructure.* Swedish Research Council.
Tindemans P (2009) Post-war research, education and innovation policy-making in Europe. In H Delanghe, U Muldur and L Soete (eds) *European science and technology policy: Towards integration or fragmentation?* Edward Elgar Publishing, pp 3–23.
Weinberg A (1961) Impact of large-scale science on the United States. *Science* **134** (3473): 161–4.
Weinberg A (1967) *Reflections on big science.* Pergamon Press.
Westfall C (2003) Rethinking big science: Modest, mezzo, grand science and the development of the Bevalac, 1971–1993. *Isis* **94**: 30–56.
Westfall C (2008) Surviving the squeeze: National laboratories in the 1970s and 1980s. *Historical Studies in the Natural Sciences* **38** (4): 475–78.

2. Methods and strategies in the study of Big Science and Research Infrastructures: A review

Nicolas Rüffin

1. INTRODUCTION

Researchers interested in the politics of Big Science and Research Infrastructures (RIs) can resort to a large body of scholarly literature. Yet this abundance of research poses a challenge, because research is scattered across different disciplines. The available studies sometimes differ fundamentally with regard to research approaches, methodologies and the set of methods employed in the investigation. This is of course also due to different research questions, theoretical assumptions and objects of study. For instance, research evaluation studies pursue approaches different from Science and Technology Studies (STS). Historiographic research differs from analyses of economic and social impact. Apart from being a source of frustration, this situation also impedes the comparative examination of cases. Moreover, the wealth of literature available does not mean that researchers automatically find appropriate methods for studying the specific aspect of the *politics* of such infrastructures and research projects.

Politics, in this context, is understood in a practical and broad sense. From the beginning, this dimension of the study of Big Science and RIs has been present among its advocates and adversaries alike, as "Big Science projects could not avoid becoming embroiled in institutional, bureaucratic, and national politics" (Capshew and Rader 1992: 13; cf. Weinberg 1962). Due to its demand for resources and subsequent need of political support, Big Science has always had an important role in the considerations of policymakers, and this role does not seem to diminish even though the political landscape and trends in its governance change (Jacob and Hallonsten 2012). The presented understanding of politics includes policy trends on the macro level, meso-level processes and micro-level politics. For instance, it subsumes changes in the general governance of science that are relevant to the sphere of Big Science and RIs, as well as the political side of the development of individual research organizations, instruments and sites, and decision-making processes (e.g. on personnel or sites) in projects and organizations. This understanding is not confined to state policies or individual actions but allows the inclusion of all aspects of the complex interactions of individual researchers, research organizations, of private and public funders and governments (Elzinga 2012). As will be shown below, studies that explicitly focus on the political side represent only one small part of the literature on Big Science and RIs (e.g. Bolliger and Griffiths, ch 5 in this volume; Cramer, ch 3 in this volume; Elzinga 2012; Hallonsten 2012; Kay 1994; Krige 2003; Papon 2004) and do not cover the full range of methods and research strategies.

Against the backdrop of an extraordinary diverse landscape of research, the main aspiration of this chapter is to provide an inventory of the field: What methods are common and particularly suitable for a given case? What research questions are associated with the selection of specific methods? Are there preferences and standards in different disciplines and research traditions? How complex and time-consuming are the methods used in the research? Are there any gaps or blind spots in the application of appropriate methods? The chapter is conceptualized, on the one hand, as an up-to-date supplement to more classical readings on Big Science (Capshew and Rader 1992; Galison and Hevly 1992). On the other hand, it also covers current studies and intends to offer a basic overview of the literature on the politics of Big Science and RIs. To this end, a corpus of studies on the topic is analyzed in terms of the research methods and strategies employed.

The chapter is structured as follows: First, the scope of the chapter is established by elaborating on the definitions of Big Science and RIs. Subsequently, this definitional groundwork constitutes the rationale for the selection of relevant studies from the literature. Section 3 contains the discussion of methods of data collection. Section 4 portraits common strategies for data analysis. Section 5 sums up the findings in order to identify trends and blind spots of research on Big Science and RIs, and to highlight opportunities and constraints for future investigations.

2. DEFINITIONS AND DATA COLLECTION

2.1 Definitions of Big Science and Research Infrastructures

Before analyzing the landscape of research on Big Science and RIs, some definitional considerations that demarcate the subject matter and scope are necessary. Several conceptual facets are already described in detail in Chapter 1 and will not be reiterated here. However, the introduction provides crucial insights for the selection of the studies in the present review. Therefore, some important findings are briefly summarized in the following.

Big Science is no one-dimensional concept, but rather an umbrella term for diverging activities and structural properties. There are several ways to categorize organizations, projects and infrastructures in a Big Science context. This includes shifting scholarly perspectives (Capshew and Rader 1992), economic and geographical differentiations (Galison 1992) or differences between disciplines (e.g. between big physics and big biology; see Vermeulen 2016). In addition, one could define Big Science according to nuances in the history of ideas (Hallonsten 2016a; Cramer et al, ch 1 in this volume), the styles of scientific work (e.g. small science on big machines; Hallonsten 2009) or classifications of scale such as Westfall's (2003) distinction of modest, mezzo and grand science or Jacob's and Hallonsten's (2012) distinction between Big Science and megascience. In recent years, the notions of Big Science have been complemented by the concept of RIs. Yet the newer term hardly serves any purpose analytically because it is at least as blurry as the older notions of Big Science. It is used to denote a multitude of different institutions that are embedded in all kinds of disciplines in the natural sciences or the humanities and social sciences. These institutions also differ massively in their size, their scope of application and involved political constellations (Hallonsten 2020; Cramer et al, ch 1 in this volume). Depending on which of these categorizations one would choose, the design of a methodically oriented review would favor accordingly selective readings. However, this chapter adopts the opposite approach by looking at the plethora of diverging without limiting itself to a particular viewpoint. In other words, the subject of the chapter is the union (as notated by $\cup$ in set theoretic terms) of the different conceptualizations in order to cover as wide a range of methods as possible.

2.2 Data Collection

The selection of the studies considered is based on a literature search in Clarivate's Web of Science (WoS), which operated with search terms such as "big science", "research infrastructures", "large-scale research infrastructures" and "megascience" and their variations. In addition, several works were added to the set that did not appear in the literature database yet are widely known and cited in the field such as the volume edited by Galison and Hevly (1992). The other chapters of the present volume were taken into account as well. In total, 152 publications were examined with regard to the methods and research strategies used; these are marked with an asterisk (*) in the reference list to this chapter. With a few exceptions, these are exclusively publications from scientific journals or books. The set includes publications from disciplines such as history of science, sociology of science, management and economics,

Table 2.1 Overview of clustered topics and associated number of studies

Topic/case	Number of studies
National laboratories of the United States	43
European facilities (except CERN)	34
CERN	26
Comparative research on laboratories	14
Field-affiliated research	15
Research infrastructures	10
Big science in general and on the national level	6
Others	4

political science and anthropology. It is not restricted to studies on European institutions but also covers a major body of literature from the United States (US). Most of the selected studies are concerned with specific cases whereas a few publications deal with the development of Big Science and RIs on a macro level (see Table 2.1).

As expected from the chosen wide-ranging selection criteria, the studies do not show convergence with regard to definitions. However, this seems to be secondary in comparison to the gain in methodological breadth, which is guaranteed by the open definition. This inclusive approach means that not every single study can be presented in the chapter. Instead, examples are mentioned in the text while a quantitative overview is provided in Table 2.2.

To navigate the jungle of diverging research designs and objects of study, the review takes an agnostic stance towards theoretical traditions and preferences of quantitative or qualitative techniques. It relies on a conventional distinction of categorizing research methods by distinguishing methods of data collection from methods and strategies of data analysis. This is, of course, an artificial distinction, since many studies follow a case study design that mixes several methods (see Tables 2.2, 2.3 and 2.4). It is purely analytical as it helps to structure the reviewed literature with regard to the methods involved.

3. METHODS OF DATA COLLECTION

The methods of data collection are presented according to their degree of reactivity. Interview techniques and participatory observation form one extreme of this scale. They establish a direct link between the investigator and the actors. At the other end of the spectrum, there are studies that rely on quantitative databases or secondary analysis of existing data and do not require that the researcher gets in contact with the investigated groups of people.

Table 2.2 *Combinations of methods in the selected publications*

	IN	PO	SU	AR	PD & OA	AUR	SD	DS & I
IN	82							
PO	6	12						
SU	1	–	5					
AR	52	2	–	68				
PD & OA	55	4	1	43	78			
AUR	–	–	–	1	–	14		
SD	6	–	–	–	2	–	19	
DS & I	11	–	1	4	10	–	14	30

Note: Based on 152 publications; values in the principal diagonal indicate absolute uses of the respective technique. Abbreviations: IN = interview, PO = participant observation, SU = survey, AR = archival research, PD & OA = public discourse and official announcements, AUR = autobiographical remarks, SD = scientometric data, DS & I = descriptive statistics and indicators.

3.1 Interviews

Interviews are among the most commonly used methods in research on Big Science and RIs (see Table 2.2) and can be found in studies on very different topics, most prominently on the emergence and development of individual laboratories and projects. Already the first comprehensive historical accounts of the European nuclear and particle physics laboratory CERN included 55 interviews with contemporary witnesses and protagonists of the founding epoch (cf. Hermann et al 1987). Another extraordinarily extensive example is the history of the US-American Fermilab near Chicago by Hoddeson et al (2008). This book is the culmination of a longer research process that was documented in detail over the years (see e.g. Westfall 1989, 2002; Westfall and Hoddeson 1996). The research team conducted more than 160 interviews during 29 years, including multiple interviews with the organization's leadership. In the case of the third director, John Peoples, multiple interviews were carried out over a period of 20 years. This extensive data set enabled the researchers to trace the laboratory's development over several decades. Interviews can also be informative supplements to other methods, such as archival research as in the case of Cramer's (2020) investigation of the European Synchrotron Radiation Facility (ESRF) and European X-Ray Free-Electron Laser (XFEL).

Interviews owe part of their popularity to their flexible areas of application. They allow covering specific issues in great detail in direct interaction with the interviewee, thus tapping into bodies of "authentic", first-person knowledge on historical processes. The application of interviews always includes subjective beliefs and viewpoints of the interviewees which means that the interviewer has to take into account post hoc justifications, strategic statements or rhetorical attempts of manipulation. Controversial statements might point to interesting debates and conflicting perspectives in the respective policy process.

However, the widespread use of the technique has a number of shortfalls, in particular a lack of rigor and reflection. From a methodological point of view, there are several types of interview methods. Klemm and Liebold (2017) distinguish hermeneutic interviews from content-oriented and ethnographic interviews. The former focuses on the interviewee's individual narration of events and processes and aims to embed and understand these stories in their social context (see also Czarniawska 1998). However, this type of interview is rarely used by scholars of Big Science and RIs, as they often conduct research on phenomena on the macro or meso level and are less interested in stories of individuals or biographical information. Ethnographically oriented interviews, on the other hand, are usually not very formalized (see Section 3.2). The majority of interviews employed in studies on Big Science and RIs take the shape of content-oriented expert interviews (for an overview, see Gläser and Laudel 2010). The function of expert interviews is to provide the researcher with facts and figures as well as insider knowledge concealed from public record.

Apart from methodological pitfalls, the application of interviews is restricted by matters of scale and access. Interviews are quite time-consuming to conduct and thus large-scale interview-based studies almost necessarily rely on teamwork. In other cases, interviews are not practicable since key actors decline interview requests or are not available, for instance, because of retirement or death. Records of interviews may sometimes be a remedy, such as in the case of the "oral history of Europe in space" program which comprises about 60 interviews with administrators, scientists and astronauts involved in the unfolding of European space policy of the twentieth century (see European University Institute 2019).

3.2 Participant Observation

The rise of laboratory studies in STS gave birth to a new form of investigation of scientific organizations, namely the ethnographic approach to Big Science (for an overview of laboratory studies, see Knorr Cetina 1995). Anthropologists and sociologists started to take part in the endeavor to look into the everyday life of natural scientists, focusing in particular on the ways scientific knowledge is produced in the laboratory (classics include Knorr Cetina 1999; Latour and Woolgar 1986). Among the different cases, Big Science organizations often were the natural starting points for ethnographic research. For instance, Knorr Cetina (1981, 1999) and Roy (2012) investigated the scientific life at CERN, Traweek (1988) based her insights on visits to the Stanford Linear

Accelerator Center (SLAC) and Doing (2009) acted as a participant observer at the Cornell High Energy Synchrotron Source.

These studies have provided us with a rich body of theories and empirical observations that massively propelled the scholarship in STS and in the sociology of knowledge in general. They present an incomparably detailed and dense view of the processes and conditions of work in RIs; and they point to the intricate relationship between the production of scientific facts in the lab and their societal – and particularly political – context, for example, by highlighting the role of, and competition for, funding (Latour and Woolgar 1986).

However, participant observations that explicitly address the governing bodies of Big Science organizations represent a blind spot in the literature. When it comes to the topic of decision-making processes in committees, advisory boards or organizational administrations, the available data are very limited. One could speculate that this is caused by a lack of attention both by STS and political science. While the former is interested in the production of knowledge, the latter is all too often committed to quantitative methodologies and analyses on the state/macro level and therefore disregards methods like participant observation (Gillespie and Michelson 2011). Apart from disciplinary habits, it is usually difficult to gain access to these bodies. After all, it is easier for policymakers to discuss matters in their ordinary fora if they do not have to fear that information is being made public by third parties, which is an argument often used to deny access to meetings and sessions of decision- and policymaking bodies of Big Science and RI organizations, in the author's own experience. What holds true for these applies to an even greater extent to political administrations. Relevant ministries and departments seldom open their gates to social scientists interested in their internal affairs and procedures. This is further complicated by the fact that multinational projects involve not just one actor, but a whole range of actors from several countries. In consequence, much of the politics of Big Science and RIs is pursued in secrecy in black-boxed circumstances. This last aspect – the matter of access and trust building – touches upon only one challenge in conducting participant observation. It is time-consuming to establish the field access, and afterwards the researcher must take care to maintain his own view on the facts and not to simply adopt viewpoints of the actors under investigation.

3.3 Survey

Surveys are relatively rare and seldom combined with other methods (see Table 2.2). Apart from some applications by economists, particularly with regard to economic impact and the relationship between Big Science and RI institutions and industry (e.g. Autio et al 2003; Florio et al 2018), applied policy research and commissioned studies also rely on surveys if they aim to determine the impact of Big Science facilities (for an overview, see Simmonds et al 2013). One can only speculate why surveys are mostly neglected as a tool in research on Big Science and RIs.

First of all, it should be noted that many surveys have the purpose to generate large data sets for quantitative analyses. However, in the field under study here, interactions often take place in small groups composed of selected scientists and policymakers. This, secondly, also points to the fact that many activities happen in relative secrecy. Surveys help little to detect and capture such hidden policy processes. Thirdly, surveys are often tied to contemporary questions and are therefore less suited to investigate historical processes.

With the advent of transformed Big Science (Hallonsten 2016a) and the increasing interest in studies on users of research facilities (D'Ippolito and Rüling 2019), scholars could revamp surveys as an adequate tool for research as, for instance, in the study by Olander (2013) who conducted a survey among users of MAX-lab in Lund, Sweden.

3.4 Documents

Along with interviews, written communication is the most extensively used data in the research on Big Science and RIs and is available in a variety of forms.

3.4.1 Archival resources

Much of the work in investigating Big Science and RIs is done in archives (see Table 2.2). This comes as no surprise given that many studies focus on the genesis of specific projects and organizations and follow a historiographic approach. While direct observation of policy processes in Big Science institutions is often not possible, organizational bureaucracies normally ensure thorough documentation. For instance, the CERN history project employed documents from at least 15 different archives in order to provide a multifaceted, monumental display of the laboratory's history (Hermann et al 1987: xix–xx). Archival resources form the basis or at least an integral part not only for the studies on CERN, but also for the history of the ESRF (Cramer 2017), the European Space Agency (Krige et al 2000), Fermilab (Hoddeson et al 2008) and others. These examples show that extensive research, and often multischolar collaboration, is necessary to find all relevant sources. This includes to identify those archives that may contain suitable material, to visit them and finally to review the documents for relevance and validity. Even though this procedure may seem laborious, it often enables a precise post hoc analysis of historical policy processes.

Apart from the efforts in searching for archival records, more basic problems stem from the political nature of many files. Documents may be classified or only made available many years after the events of interest; for instance, the archives of the European Molecular Biology Laboratory (EMBL) and the European Southern Observatory (ESO) are not open for social research, as the author was recently informed. In addition, many archival documents have undergone an editorial process and thus only represent a specific segment of the political dimension of Big Science and RIs. This limitation can sometimes be compensated for if archives also hold personal records of individuals who were involved in the processes under investigation.

3.4.2 Public discourse and official announcements

The discourse on Big Science and RIs does not exclusively take place behind closed doors. The governance of the organizations produces large amounts of documents, such as conventions, rules of procedure and additional legal documents like bilateral or multilateral treaties, procurement contracts, press releases, brochures, annual reports and policy papers. One example is Franssen's (ch 7 in this volume) use of policy reports from various actors in the Dutch science policy landscape to track the steps of research infrastructure development in the humanities. But the diverse opportunities for data collection from publicly available material become even more visible in the example of a single Big Science facility, the European Spallation Source (ESS) in Lund, Sweden. Yu and colleagues (2017) delve into the juridical aspects of intellectual property rights rules within the European Research Infrastructure Consortium (ERIC) framework based on public sources and administrative decisions. Stenborg and Klintman (2012), on the other hand, rely on articles from newspapers to sketch the local (non-)resistance against the very same new infrastructure. Agrell (2012) refers to official documents issued by the ESS administration to study the promotion campaign to locate the facility in Lund. Finally, Liljefors (2013) extends the range of possible material even further by analyzing videos uploaded on YouTube to reconstruct the discourse on the ESS. The plethora of publicly available documents is both a challenge and an opportunity. The analytical results of different data sets make it possible to draw a holistic picture, yet this may also lead to contradictions in the narration and a lack of comparability between studies.

3.4.3 Autobiographical remarks

Some individuals – often after retirement – put pen to paper to reflect on their time in Big Science and RI contexts. These sources include for example publications by directors general Schopper (2009, CERN), Smith (2007, CERN) and Blaauw (1991, ESO), or former president of ESO's Council van der Kruit (2006). Other examples are the memoirs of the French diplomat de Rose (2014) who acted as a member of CERN's Council or the founding director of SLAC Wolfgang Panofsky's publications (Panofsky 1992; Panofsky and Deken 2007) on the genesis of SLAC.

These books and articles mainly present personal views and do not claim the status of research literature. Nevertheless, they represent an important source for the reconstruction of critical events within research organizations. The distinctive feature of these documents is their unique insight into the internal processes of organizations and political constellations as they are written by key personnel with access to crucial and often restricted information. Yet the same concerns apply to them as to expert interviews. Usually, these accounts are apologetic recollections of past events. It is advisable to cross-check the insights provided in this literature with additional sources and methods.

3.5 Quantitative Data

3.5.1 Scientometric databases

Big science (non-capitalized, see ch 1 in this volume) and bibliometric methods have been closely linked from their beginnings. De Solla Price, one of the founding fathers of scientometrics, employed quantitative methods to study the emergence of big science as a social and scientific phenomenon (Capshew and Rader 1992). His seminal book *Little Science, Big Science* (1963) looked into the transformation of scientific work in the twentieth century and paved the way for many other applications of scientometrics. Although Price viewed the trends associated with big science primarily from an analytical standpoint, he also contemplated on their political consequences.

In a narrow sense, scientometric methods deal with the counting of publications, the visualization of citations and co-publication relationships. Bibliometric analyses are mostly encountered when it comes to taking stock of developments of scientific fields (i.e. big science, non-capitalized) or to trace the performance of scientists and research organizations (i.e. Big Science, capitalized). Both applications are well connected to political considerations of economic and societal impact, return on investment and research evaluation. The latter use skyrocketed during the last decades, resulting in many studies on the performance of researchers, not least in Big Science and RI institutions.

The basis for such analyses is often formed by scientometric databases such as Clarivate's WoS. These commercial repositories contain information on publications from many scientific fields which renders them frequently used starting points for carrying out studies on Big Science and RIs. For instance, Zhang et al (2011) employed data from the WoS to evaluate the impact of the Sloan Digital Sky Survey, a large-scale astronomy project, on scientists' subsequent research. Lauto and Valentin (2013) utilized the same data source in their study of the Oak Ridge National Laboratory in Tennessee, US to investigate the facility's connections to the global science community. Relying on existing databases entails, however, certain challenges. Databases like Scopus and WoS tend to provide a better coverage of journals from English-speaking countries and have limitations in the periods covered (Mongeon and Paul-Hus 2016). More importantly, these databases do not systematically and reliably record the affiliation of authors with regard to RIs and data. It might prove difficult to identify the connections between researchers and infrastructures or organizations on the basis of the unedited records, in particular since user facilities (Hallonsten 2016b) often welcome researchers on a temporary or virtual basis who are otherwise affiliated to universities. The usually no less time-consuming alternative to these databases is to independently develop a case-oriented original data set. This can be achieved by searching relevant journals (Martin and Irvine 1984a, 1984b), official reports issued by Big Science organizations (Hallonsten 2013), internal organizational databases (Canals et al 2017) or online sources (Heidler and Hallonsten 2015). However, no single source can guarantee achieving exhaustive coverage.

3.5.2 Descriptive statistics and indicators

It has been pointed out several times that scientometric data represent only a small proportion of possibly relevant numbers with regard to Big Science (Hallonsten 2013, 2014; Heidler and Hallonsten 2015). De Solla Price's book (1963) already made use of simple indicators like, for example, the number of scientists or universities founded in Europe between the Medieval and modern times; subsequent studies followed suit in combining genuine bibliometric and non-bibliometric indicators (e.g. Irvine and Martin 1984), "often designed for answering particular research questions" (Moed et al 2004: 11). In particular, indicators are a frequently encountered feature of studies from economics that investigate the impact of Big Science facilities and RIs (for a general list of science and technology indicators, see Barré 2004). Bianco and colleagues (2017) looked at the scientific performance of the International Space Station and suggested alternative indicators such as provisional patents, licenses or follow-on funding. A commissioned review by Simmonds and colleagues (2013) lists 41 studies that employed a wide range of indicators like employment data, public statistics in general, data on budgets and expenditures, and user statistics. This review also points to the fact that many reports do not specify how exactly they use these statistics, not to mention that many of the listed studies did not undergo a formal peer review.

There are, however, also examples of indicators apart from econometric studies. For instance, Heck (2000) relies on a directory of astronomical organizations to map the contemporary shape of the field. Hallonsten and Heinze (2012, 2015) illustrate in their studies on the development of synchrotron radiation facilities and national laboratories in the US and Germany how applying a "quantitative phenomenology" (Krücken et al 2010: 5), supported by plausi-

Table 2.3 Number of studies broken down by number of methods used

	One method	Two methods	Three methods	Four methods
Number of studies	49	54	45	4

ble theorizing, can build a strong argument. In these papers, the authors count the number and purpose of new facilities to show how the field develops in line with sociological theories. For suchlike studies, researchers can construct indicators from ready-made databases maintained by organizations like the Organisation for Economic Co-operation and Development (OECD), Eurostat or the United Nations Educational, Scientific and Cultural Organization (UNESCO). Alternatively, scholars have to construct new indicators from case-based material such as annual reports or other documents. In principle, the same rules of procedure and precautions apply as for scientometric data: existing databases offer easily accessible data, but may have to be cleaned and cross-checked for flaws and biases. Creating a genuine database is more time-consuming, but allows for customization.

4. STRATEGIES OF DATA ANALYSIS

There are certain combinations of methods that occur frequently in the literature (Table 2.3). The vast majority of studies employ a case-oriented view, but depending on the scholars' disciplinary backgrounds and focus of the study, there are different approaches to process the data. Basically, most of the existing literature can be divided into clusters that follow either narrative or nomothetic research strategies.

4.1 Narrative Approaches

A large part of the existing studies on Big Science and RIs can be characterized as following a narrative approach. This body of research does not rely on quantitative hypothesis testing, but wants to persuade by a convincing argumentation and the richness of material employed. However, the exact procedures differ between historiographic and ethnographic studies.

4.1.1 Historiography

Historical studies represent a classical approach to research on Big Science and RIs. This can be seen from the fact that not only genuine historians of science (Hermann et al 1987; Hoddeson et al 2008) are committed to this strategy, but also that many studies with a social science orientation concentrate on it (Cramer 2017; Hallonsten 2009). Regardless of the data employed, this

Table 2.4 Five combinations of methods most usually employed in the data set

Methods applied	Number of studies
Interview, archival research, public discourse and official announcements	33
Interview and archival research	15
Autobiographical remarks	13
Interview, public discourse and official announcements	13
Scientometric data, descriptive statistics and indicators	9

Note: Overall, there were 28 different combinations of methods of data collection.

approach aims to describe and explain how the founding and development of individual organizations or fields has taken place, or at least to present the different perspectives on these processes. Usually, the research is structured chronologically by reconstructing the events and contextual conditions that led to the founding of a laboratory (e.g. Cassata 2015; Hermann et al 1987; Krige 2002; Strasser 2003), the advent of new projects or infrastructures (Maxson Jones et al 2018; Moskovko, ch 6 in this volume) or the emergence of a new field of research (Hallonsten and Heinze 2015; Vermeulen 2016). Some social scientists aim to study the macro level of (European) science and technology policy and trace all the aforementioned developments at once (e.g. Elzinga 2012; Nedeva 2013; Papon 2004). Big Science and RIs play a central role in these overviews. Krige in particular attempted to integrate the findings from the historical studies on intergovernmental Big Science organizations in which he was involved into a comprehensive narrative structure (Krige 2003).

Scholars in this tradition for the most part combine the trinity of interviews, archival resources and public documents (see Table 2.4). As most of the events under investigation took place long ago, participant observation is often not a viable option in these instances. The use of indicators and statistics is also considered in a supportive manner, provided that it serves the purpose of argumentation. Heinze et al (2017), for instance, use statistical data on the science budget in Germany to bolster their argument of external pressures that led physicists to explore new technological and organizational trajectories, in particular synchrotron radiation.

4.1.2 Ethnography

In contrast to historiographic studies, research inspired by ethnographic tradition takes a different route. It is focused on developing what the anthropologist Clifford Geertz (1973) called "thick descriptions", that is a comprehensive observation of the behavior of the investigated population *and* the contextual factors that play into this behavior.

This is associated with a decoupling from a narrowly defined canon of methods. Instead, ethnographers of Big Science prefer a more flexible processing of data in which the collection method of participant observation can hardly be overestimated. To a certain extent, it forms the umbrella under which many other methods such as interviews, the analysis of official documents or more informal writings like, for instance, laboratory logs can be integrated. The approach is holistic and often explorative. A narrative that combines all impressions and observations from the field normally emerges in the course of the investigation. Thus, there are not only connections to Geertz's anthropology, but also to the approach of Glaser and Strauss's (1967) *Grounded Theory Methodology*, that primarily aims at developing new theoretical insights.

For example, in his study on the laboratories at Cornell University, Doing (2009) transforms all of his observations into three separate stories that he subsequently links argumentatively to describe the organizational changes from particle physics to biological research. The display of his observations takes up the same space as his reflection of this data collection. Thus, it shows very well the "unfiltered" perspective of an ethnographic approach that rejects the narrowing to a few, especially quantitative, methods.

4.2 Nomothetic Approaches: Descriptive and Inferential Statistics

Nomothetical research on Big Science and RIs relies on data and instruments from quantitative social research to calculate and identify recurring patterns and law-like properties. These methods of data analysis relate to the classical repertoire of scientometrics and econometrics. Econometrics and innovation studies produce the lion's share of nomothetic studies as these fields tend to prefer quantitative methods over qualitative approaches. These include techniques like correlations and factor analyses (Autio et al 2003) as well as regression models (Lauto and Valentin 2013). Scientometric data have traditionally been of particular interest to many scholars of big science. De Solla Price (1963) employed descriptive statistics to track the growth of the sciences since the seventeenth century. In addition, he used the mathematical knowledge of logistical growth to concisely bolster his argument.

Subsequent publications have continued to make use of the descriptive scientometric indicators discussed in Section 3. For example, Irvine and Martin (1984) chose a descriptive approach in their evaluation of CERN's performance. While a mere description would not yet be very informative, their study gained strength by comparing the infrastructures in Geneva with the performance of other particle physics laboratories in a standardized way. In a similar manner, Davidse and Van Raan (1997) compared CERN, the German Electron Synchrotron (Deutsches Elektronen-Synchrotron, DESY), and SLAC with regard to their recognition by applied research and development. Since de

Solla Price's pioneering publications, scientometrics has advanced considerably in terms of applied methods, resulting in ever more sophisticated statistical procedures. Zhang et al (2011) give a good example in their study of the patterns of collaboration associated with the Sloan Digital Sky Survey. They rely on the statistical concept of *entropy* to track the development of research topics among astronomers.

4.3 Network Analysis

The instruments of network analysis constitute an essential part of the toolbox of social scientists (e.g. Scott 2000; Wasserman and Faust 1994). Network analyses often are employed for explorative research, i.e. to study, for instance, the patterns of (international) collaboration, regularities in the development of scientific communities or social structures in organizational fields.

Scholarship on networks has produced various metrics to describe these complex structures. These include parameters like, for instance, the *density* (i.e. the number and ratios of connections within the network), the *degree* of individual nodes within a network that characterize, among others, the centrality of scholars and policymakers within a given constellation, or the clusters and subgroups in a broader network (for a detailed presentation of the individual indicators, see Wasserman and Faust 1994). The study of Canals et al. (2017) on cooperation within the ATLAS project at CERN is an exemplary application of these techniques. The authors analyze the institutional co-authorship networks within this particular project and identify four communities of roughly the same size. They use these results from network analysis to support their argument that ATLAS does not fit into the conventional conception of scientific research collaborations but represents a different, megascience-based type of research. In a similar manner, Lozano et al (2014) investigate the scientific work at the archaeological digging site of Atapuerca, Spain, by means of co-authorship networks.

However, these applications of network analysis almost always focus exclusively on scientific networks which already points to one of the underlying deficits in Big Science and RI scholarship which subsequently will be discussed in more detail.

5. DISCUSSION

The current situation of research on Big Science and RIs is characterized by a high diversity of approaches that cluster into multiple research traditions. While several studies succeed in combining methods and theory to shed new light on Big Science and RIs, there remains a need to balance between in-depth, tailor-made case studies and systematic comparative analyses in order to identify the special features of individual cases vis-à-vis the general characteristics of the politics of Big Science and RIs. This objective requires the application of sound social theorizing and sophisticated research designs.

5.1 Strengths and Blind Spots of Current State of the Art

The review shows that the combination of interviews and document analysis – sometimes supplemented by selected quantitative indicators or participant observation – has established itself as a widespread standard for narrative investigations of Big Science institutions and RIs. The works of historians of science like Hermann et al (1987) or Hoddeson et al (2008) are exemplary of a very high standard of case-oriented research. At the same time, the use of indicators and quantitative methods has developed significantly within the communities of scientometrics and economics since their beginnings in the 1960s. Applying sophisticated statistical methods, studies from this strand of research have contributed immensely to our understanding of collaboration and scientific development in the field. Despite the high standards in many areas of research, the aim should always be to review and, if necessary, improve current practices. A systematic and transparent application of methods is essential in this endeavor. However, when looking at the body of literature, there are at least three major gaps or blind spots.

The first emerges within the framework of narrative approaches. In historiographic research, the policy process remains a black box that is only illuminated post hoc once restrictions on archival resources are lifted or interview partners are retired. Ethnographic studies, on the other hand, fail to go beyond the laboratory and do not directly study the actors from the political system. While there are plenty of studies on the everyday life of scientists, starting with classics like Latour and Woolgar (1986) to newer examples like ethnographies by Roy (2012) or Doing (2009), investigations of national science administrations and other political institutions concerned with Big Science and RIs are scarce due to the principles of secrecy and restricted access that surround governmental activities. Yet examples from other policy areas illustrate that participant observations of policymakers can yield new and valuable insights (Neumann 2012; Pritzlaff and Nullmeier 2011; see also Gläser and Laudel 2016). The second gap arises between nomothetic and narrative approaches. Despite some exceptions from the rule, there are seldom systematic triangulations of quantitative and qualitative data and analytical instruments among the studies of Big Science and RIs (see Table 2.2; notable exceptions are Martin and Irvine 1984a, 1984b). In general, scientometrics seems to be a separate community dealing with the description and explanation of scientific communication, collaboration or impact. This community has developed a range of highly standardized indicators that help to capture the scientific work in Big

Science and RI contexts and measure the (potential) added value of these institutions. But there is little cross-fertilization with research that follows a narrative approach. The separation seems to be based on disciplinary traditions and contradicting philosophical and theoretical worldviews. Nevertheless, it is an open question whether – and how – more sophisticated methods of quantitative research can be combined with narrative-oriented methods in order to gain new insights and avoid blind spots with regard to the policy process. Finally, intensive reflections on the methods used are rare. There are also seldom any considerations as to how the canon of methods can be improved and extended. Methods of interview analysis, for instance, are not very sophisticated and systematic in terms of broader sociological methodologies and theories. The widespread use of expert interviews is limited to generating content that complements the narration of the respective study. Frequently, no statement is made as to which methods are used for analysis, how a systematic evaluation was carried out and which basic methodological assumptions are used. In sum, the question remains as to whether creative combinations of methods help to illuminate blind spots and black boxes.

5.2 Prospects and Challenges for Future Research

Against the backdrop of the analysis, this section contains two remarks about the routes that research on Big Science and RIs, and in particular its political dimension, could take in the future and the role that methods play in this endeavor.

The first approach consists of new and innovative combinations of established methods. There are some examples that a combination of scientometric and qualitative methods can help to illuminate the features of a particular case *and* its place within a wider macro context of political bargaining and scientific progress (e.g. Martin and Irvine 1984a). One could, for example, investigate the extent to which patterns of scientific and political communication are related to organizational changes in the policy area of Big Science and RIs. In terms of methods, this would mean combining scientometric, discourse-analytical and (qualitative) organizational analyses. A research question attached to this comprehensive approach could be whether increasing transnational scientific communication is responsible for the founding of international laboratories. In other words, this is the question of whether examples like CERN and the EMBL have set a precedent. The foundation of these organizations was significantly driven by the exchange between physicists and biologists from different countries (Hermann et al 1987; Krige 2002). There are certainly also cases in European history in which the processes went exactly the other way around, i.e. that policymakers considered setting up Big Science institutions and RIs and only after that scientists jumped on the bandwagon. The rise of infrastructural policies in the European Union might be a case in point for this order of events (cf. Hallonsten 2020; Ulnicane, ch 4 in this volume). Another venue for combined methods could be to enhance narrative approaches via the use of quantitative methods that bolster the claims made in the argument. For instance, Pestre and Krige (1992) prominently made a case that the early history of CERN was decisively shaped by a network of a few scientists and science policymakers. It would be worthwhile to investigate, for instance, by using network analysis based on data obtained by qualitative methods like interviews and archival material, whether similar structures emerge(d) in other instances.

The second direction to further the development of methods is to improve the individual tools and their application. This means using individual instruments more systematically, but also thinking about applying them to new questions and fields. In the case of qualitative methods, researchers should be more concerned with what methods are available that go beyond the mere elicitation of factual knowledge. This could mean, for example, using the Grounded Theory Methodology to open up new perspectives on Big Science and RIs or to look at interview data from the perspective of biographical research in systematic, transparent ways (for an overview, see Rosenthal 2018). As already mentioned above, this route for improvement also includes conducting participant observation in unprecedented locations and situations such as ministries of science and technology or the administration of Big Science and RI organizations as it is a viable question how ministerial officers' backgrounds and worldviews play into their understanding and subsequent handling of Big Science and RIs (Neumann 2012; Pritzlaff and Nullmeier 2011). Delving into this topic could help to better understand how administrative problems are solved and how decisions are made. The topic of policy networks is another case in point. Up until now, the extensive analysis of scientific networks is not matched by a corresponding analysis of the science and technology policy networks, at least with regard to Big Science and RIs (see Grande and Peschke 1999). Some researchers have highlighted the role of international organizations like, for instance, the OECD in the formulation of national research policies (e.g. Henriques and Larédo 2013). Applying techniques from network analysis as well as participant observation might yield new insights as to whether these diffusion processes correspond with increasing links between policymakers of Big Science and RIs in Europe.

Quantitative methods need large numbers of data to conduct their analyses and produce reasonable results. It is therefore an important question how to obtain such data from sources different from scientometric databases. Except from small-scale processes that involve just a few actors, other questions could be answered by the quantitative analysis of archive documents. The use of these data sources might subsequently also lead to the creation of new indicators to characterize (and benchmark) developments in the policy area. In this regard, attention should be paid to the rise of big data and automatic analyses of large-scale sets of data. Zhang et al (2011) already illustrated how automatic analytical methods may shake up and amend our understanding of scientometrics. The data available in the big databases, but also in the archives of ministries and Big Science organizations and RIs, might provide an excellent venue for testing advanced automatic methods. Furthermore, a standardization of methods would help to establish comparability between different phenomena. Standardized techniques could be employed to systematically either support or challenge narrative accounts such as those by Krige (2003) or Papon (2004) that rely on comparative, yet small samples.

Finally, it would be worthwhile to turn to newer cases to test established and new methods. There are a number of interesting cases (see Ulnicane, ch 4 in this volume; Moskovko, ch 6 in this volume; Franssen, ch 7 in this volume; Hallonsten 2020). In the field of synchrotron radiation alone, there are examples like SESAME, the first intergovernmental research organization in the Middle East, the Extreme Light Infrastructure (ELI) with three sites in Hungary, Romania and the Czech Republic, or envisaged projects like the African Light Source that wait to be explored by social scientists. In particle physics, it will be very interesting to observe the policy processes tied to the next generation of accelerators that could succeed the Large Hadron Collider at CERN. Currently, there are a number of candidate infrastructures planned in Switzerland, Japan and China (Castelvecchi 2019). Whether these gigantic machines will ever be built, however, is questionable due to the costs and organizational problems associated with these megaprojects. Actually, these instances point towards another important aspect of the development of Big Science and RIs: the need for adaptation and planning when the projects' original missions phase out. Hallonsten and Heinze (2012, 2015) as well as D'Ippolito and Rüling (ch 11 in this volume) have presented valuable insights into such processes, yet it would be interesting to put them to the test using other projects and facilities. It should also not be forgotten that the concept of infrastructure also covers projects beyond the natural sciences. As Franssen (ch 7 in this volume) pointed out, there is also a considerable momentum in the formation of RIs in the humanities. Beyond these fields there are even more cases, such as the growing number of ERICs in Europe, the ongoing global expansion of established intergovernmental research organizations and the initiatives to found a new European organization dedicated to the study and advancement of artificial intelligence research. As these examples are more recent, they might provide opportunities for the study of political actors outlined above. Without question, more RIs will come into being in the future, leaving the scholar of Big Science and RIs with countless chances to conduct new, innovative and methodologically sound research.

ACKNOWLEDGMENTS

The author would like to thank the editors and Thomas Franssen for helpful comments and suggestions on earlier drafts of the manuscript that greatly helped to improve the chapter.

REFERENCES

Åberg S and A Bengtson (2015) Does CERN procurement result in innovation? *Innovation: The European Journal of Social Science Research* **28** (3): 360–83.*

Agrell W (2012) Framing prospects and risk in the public promotion of ESS Scandinavia. *Science and Public Policy* **39** (4): 429–38.*

Andersen P H and S Åberg (2017) Big-science organizations as lead users: A case study of CERN. *Competition and Change* **21** (5): 345–63.*

Aronova E, K S Baker and N Oreskes (2010) Big Science and Big Data in biology: From the International Geophysical Year through the International Biological Program to the Long Term Ecological Research (LTER) Network, 1957–present. *Historical Studies in the Natural Sciences* **40** (2): 183–224.*

Autio E, M Bianchi-Streit and A-P Hameri (2003) *Technology transfer and technological learning through CERN's procurement activity*. CERN Education and Technology Transfer Division.*

Autio E, A-P Hameri and M Nordberg (1996) A framework of motivations for industry–big science collaboration: A case study. *Journal of Engineering and Technology Management* **13** (3–4): 301–14.*

Autio E, A-P Hameri and O Vuola (2004) A framework of industrial knowledge spillovers in big-science centers. *Research Policy* **33** (1): 107–26.*

Baneke D (2019) Let's not talk about science: The normalization of Big Science and the moral economy of modern astronomy. *Science, Technology, and Human Values* **45** (1): 164–94.*

Barré R (2004) S&T indicators for policy making in a changing science–society relationship. In H F Moed, W Glänzel and U Schmoch (eds) *Handbook of quantitative science and technology research: The use of publication and patent statistics in studies of S&T systems*. Kluwer Academic Publishers, pp 115–31.

Bianco W, D Gerhart and S Nicolson-Crotty (2017) Waypoints for evaluating Big Science. *Social Science Quarterly* **98** (4): 1144–50.*

Blaauw A (1991) *ESO's early history: The European Southern Observatory from concept to reality*. ESO.*

Bodnarczuk M and L Hoddeson (2008) Megascience in particle physics: The birth of an experiment string at Fermilab. *Historical Studies in the Natural Sciences* **38** (4): 508–34.*

Boyack K W and N Rahal (2005) Evaluation of laboratory directed research and development investment areas at Sandia. *Technological Forecasting and Social Change* **72** (9): 1122–36.*

Canals A E, M Ortoll and M Nordberg (2017) Collaboration networks in big science: The ATLAS experiment at CERN. *El Profesional de La Información* **26** (5): 961–71.*

Capshew J H and K A Rader (1992) Big Science: Price to the present. *Osiris* **7**: 2–25.

Cassata F (2015) "A cold spring harbor in Europe": EURATOM, UNESCO and the foundation of EMBO. *Journal of the History of Biology* **48** (4): 539–73.*
Castelvecchi D (2019) Next-generation LHC: CERN lays out plans for €21-billion supercollider. *Nature* **565** (7740): 410.
Cogen M (2012) Membership, associate membership and pre-accession arrangements of CERN, ESO, ESA, and EUMETSAT. *International Organizations Law Review* **9** (1): 145–79.*
Cole S and G S Meyer (1985) Little science, big science revisited. *Scientometrics* **7** (3–6): 443–58.*
Cramer K C (2017) Lightening Europe: Establishing the European Synchrotron Radiation Facility (ESRF). *History and Technology* **33** (4): 396–427.*
Cramer K C (2020) *A political history of Big Science: The other Europe*. Palgrave Macmillan.*
Crease R P (2001) Anxious history: The High Flux Beam Reactor and Brookhaven National Laboratory. *Historical Studies in the Physical and Biological Sciences* **32** (1): 41–56.*
Crease R P (2005a) Quenched! The ISABELLE saga, I. *Physics in Perspective* **7** (3): 330–76.*
Crease R P (2005b) Quenched! The ISABELLE saga, II. *Physics in Perspective* **7** (4): 404–52.*
Crease R P (2008a) Recombinant science: The birth of the Relativistic Heavy Ion Collider (RHIC). *Historical Studies in the Natural Sciences* **38** (4): 535–68.*
Crease R P (2008b) The National Synchrotron Light Source, Part I: Bright idea. *Physics in Perspective* **10** (4): 438–67.*
Crease R P (2009) The National Synchrotron Light Source, Part II: The bakeout. *Physics in Perspective* **11** (1): 15–45.*
Czarniawska B (1998) *A narrative approach to organization studies*. SAGE.
D'Ippolito B and C C Rüling (2019) Research collaboration in large scale research infrastructures: Collaboration types and policy implications. *Research Policy* **48** (5): 1282–96.*
Davidse R J and A F J van Raan (1997) Out of particles: Impact of CERN, DESY and SLAC research to fields other than physics. *Scientometrics* **40** (2): 171–93.*
Deák C and I Szabó (2016) Assessing cooperation between industry and research infrastructure in Hungary. *Technology Innovation Management Review* **6** (7): 13–20.*
Del Bo C F (2016) The rate of return to investment in R&D: The case of research infrastructures. *Technological Forecasting and Social Change* **112**: 26–37.*
Doing P (2009) *Velvet revolution at the Synchrotron: Biology, physics, and change in science*. MIT Press.*
Eggleton D C (2017) *Examining the relationship between leadership and megascience projects*. PhD thesis. Available at http://sro.sussex.ac.uk/71281/ (last accessed December 2, 2019).*
Elzinga A (2012) Features of the current science policy regime: Viewed in historical perspective. *Science and Public Policy* **39** (4): 416–28.
European University Institute (2019) Oral history of Europe in space. Available at https://archives.eui.eu/en/oral_history/#ESA (last accessed December 9, 2019).
Everitt C W F (1992) Background to history: The transition from little physics to big physics in the Gravity Probe B Relativity Gyroscope Program. In P Galison and B W Hevly (eds) *Big Science: The growth of large-scale research*. Stanford University Press, pp 212–35.*

Florio M, F Giffoni, A Giunta and E Sirtori (2018) Big Science, learning, and innovation: Evidence from CERN procurement. *Industrial and Corporate Change* **27** (5): 915–36.*

Galison P (1992) The many faces of Big Science. In P Galison and B W Hevly (eds) *Big Science: The growth of large-scale research.* Stanford University Press, pp 1–17.

Galison P and B W Hevly (eds) (1992) *Big Science: The growth of large-scale research.* Stanford University Press.

Galison P, B W Hevly and R Lowen (1992) Controlling the monster: Stanford and the growth of physics research, 1935–1962. In P Galison and B W Hevly (eds) *Big Science: The growth of large-scale research.* Stanford University Press, pp 46–77.*

Geertz C (1973) *The interpretation of cultures: Selected essays.* Basic Books.

Gillespie A and M R Michelson (2011) Participant observation and the political scientist: Possibilities, priorities, and practicalities. *PS: Political Science and Politics* **44** (2): 261–5.

Glaser B F and A L Strauss (1967) *The discovery of grounded theory: Strategies for qualitative research.* Aldine.

Gläser J and G Laudel (2010) *Experteninterviews und qualitative Inhaltsanalyse als Instrumente rekonstruierender Untersuchungen*, 4th ed. VS Verlag.

Gläser J and G Laudel (2016) Governing science. *European Journal of Sociology* **57** (1): 117–68.

Granberg A (2012) The ESS project as a generator of conflict and collaboration: An assessment of the official picture of costs and benefits and the research-community response. In O Hallonsten (ed) *In pursuit of a promise: Perspectives on the political process to establish the European Spallation Source (ESS) in Lund, Sweden.* Arkiv Academic Press, pp 109–57.*

Grande E and A Peschke (1999) Transnational cooperation and policy networks in European science policy-making. *Research Policy* **28** (1): 43–61.

Gribbe J and O Hallonsten (2017) The emergence and growth of materials science in Swedish universities. *Historical Studies in the Natural Sciences* **47** (4): 459–93.*

Habfast C (2010) The DESY golden jubilee in Hamburg: Lessons from the past. *Physics in Perspective* **12** (2): 219–30.*

Hallonsten O (2009): *Small science on big machines: Politics and practices of synchrotron radiation laboratories.* PhD thesis, Lund University.

Hallonsten O (2011) Growing Big Science in a small country: MAX-lab and the Swedish research policy system. *Historical Studies in the Natural Sciences* **41** (2): 179–215.*

Hallonsten O (2012) Continuity and change in the politics of European scientific collaboration. *Journal of Contemporary European Research* **8** (3): 300–19.

Hallonsten O (2013) Introducing "facilitymetrics": A first review and analysis of commonly used measures of scientific leadership among synchrotron radiation facilities worldwide. *Scientometrics* **96** (2): 497–513.*

Hallonsten O (2014) How expensive is Big Science? Consequences of using simple publication counts in performance assessment of large scientific facilities. *Scientometrics* **100** (2): 483–96.*

Hallonsten O (2015a) The parasites: Synchrotron radiation at SLAC, 1972–1992. *Historical Studies in the Natural Sciences* **45** (2): 217–72.*

Hallonsten O (2015b) Unpreparedness and risk in Big Science policy: Sweden and the European Spallation Source. *Science and Public Policy* **42** (3): 415–26.*

Hallonsten O (2016a) *Big Science transformed: Science, politics and organization in Europe and the United States.* Palgrave Macmillan.

Hallonsten O (2016b) Use and productivity of contemporary, multidisciplinary Big Science. *Research Evaluation* **25** (4): 486–95.*
Hallonsten O (2020) Research infrastructures in Europe: The hype and the field. *European Review* **28** (4): 617–35.
Hallonsten O and O Christensson (2017) Collaborative technological innovation in an academic, user-oriented Big Science facility. *Industry and Higher Education* **31** (6): 399–408.*
Hallonsten O and T Heinze (2012) Institutional persistence through gradual organizational adaptation: Analysis of national laboratories in the USA and Germany. *Science and Public Policy* **39** (4): 450–63.*
Hallonsten O and T Heinze (2015) Formation and expansion of a new organizational field in experimental science. *Science and Public Policy* **42** (6): 841–54.*
Heck A (2000) Astronomy-related organizations: Geographical distributions, ages and sizes. In A Heck (ed) *Organizations and strategies in astronomy*. Springer Netherlands, pp 7–66.*
Heidler R and O Hallonsten (2015) Qualifying the performance evaluation of Big Science beyond productivity, impact and costs. *Scientometrics* **104** (1): 295–312.*
Heinze T and O Hallonsten (2017) The reinvention of the SLAC National Accelerator Laboratory, 1992–2012. *History and Technology* **33** (3): 300–32.*
Heinze T, O Hallonsten and S Heinecke (2015a) From periphery to center: Synchrotron radiation at DESY, Part I: 1962–1977. *Historical Studies in the Natural Sciences* **45** (3): 447–92.*
Heinze T, O Hallonsten and S Heinecke (2015b) From periphery to center: Synchrotron radiation at DESY, Part II: 1977–1993. *Historical Studies in the Natural Sciences* **45** (4): 513–48.*
Heinze T, O Hallonsten and S Heinecke (2017) Turning the ship: The transformation of DESY, 1993–2009. *Physics in Perspective* **19** (4): 424–51.*
Henriques L and P Larédo (2013) Policy-making in science policy: The "OECD model" unveiled. *Research Policy* **42** (3): 801–16.*
Hermann A, L Belloni, U Mersits, D Pestre and J Krige (eds) (1987) *History of CERN, Volume 1: Launching the European Organization for Nuclear Research*. North-Holland.*
Hermann A, L Weiss, D Pestre, U Mersits and J Krige (eds) (1990) *History of CERN, Volume 2: Building and running the laboratory, 1954–1965*. North-Holland.*
Hoddeson L (1992) Mission change in the large laboratory: The Los Alamos Implosion Program, 1943–1945. In P Galison and B W Hevly (eds) *Big Science: The growth of large-scale research*. Stanford University Press, pp 265–89.*
Hoddeson L and A W Kolb (2000) The superconducting super collider's frontier outpost, 1983–1988. *Minerva* **38** (3): 271–310.*
Hoddeson L, A W Kolb and C Westfall (2008) *Fermilab: Physics, the frontier, and megascience*. University of Chicago Press.*
Hofer F (2005) Dissemination of CERN's technology transfer: Added value from regional transfer agents. *Industry and Higher Education* **19** (4): 315–24.*
Höne K E and J Kurbalija (2018) Accelerating basic science in an intergovernmental framework: Learning from CERN's science diplomacy. *Global Policy* **9**: 67–72.*
Hounshell D A (1992) Du Pont and the management of large-scale research and development. In P Galison and B W Hevly (eds) *Big Science: The growth of large-scale research*. Stanford University Press, pp 236–61.*
Irvine J and B R Martin (1984) CERN: Past performance and future prospects, II: The scientific performance of the CERN accelerators. *Research Policy* **13** (5): 247–84.*

Irvine J and B R Martin (1985) Basic research in the East and West: A comparison of the scientific performance of high-energy physics accelerators. *Social Studies of Science* **15** (2): 293–341.*

Jacob M and O Hallonsten (2012) The persistence of big science and megascience in research and innovation policy. *Science and Public Policy* **39** (4): 411–15.

Kaiserfeld T (2013) The ESS from neutron gap to global strategy. In T Kaiserfeld and T O'Dell (eds) *Legitimizing ESS: Big Science as a collaboration across boundaries*. Nordic Academic Press, pp 25–41.*

Kay W D (1994) Democracy and super technologies: The politics of the space shuttle and space station freedom. *Science, Technology, and Human Values* **19** (2): 131–51.

Kevles D J (1992) K1K2: Korea, science, and the state. In P Galison and B W Hevly (eds) *Big Science: The growth of large-scale research*. Stanford University Press, pp 312–33.*

Klemm M and R Liebold (2017) Qualitative interviews in der organisationsforschung. In S Liebig, W Matiaske and S Rosenbohm (eds) *Handbuch empirische Organisationsforschung*. Springer Gabler, pp 299–324.

Knorr Cetina K (1981) *The manufacture of knowledge: An essay on the constructivist and contextual nature of science*. Pergamon Press.

Knorr Cetina K (1995) Laboratory studies: The cultural approach to the study of science. In S Jasanoff, G Markle, J Peterson and T Pinch (ed) *Handbook of science and technology studies*. SAGE, pp 140–66.

Knorr Cetina K (1999) *Epistemic cultures: How the sciences make knowledge*. Harvard University Press.

Kolb A and L Hoddeson (1993) The mirage of the "world accelerator for world peace" and the origins of the SSC, 1953–1983. *Historical Studies in the Physical and Biological Sciences* **24** (1): 101–24.*

Krige J (1993) Some Socio-historical aspects of multinational collaborations in high-energy physics at CERN between 1975 and 1985. In E Crawford, T Shinn and S Sörlin (eds) *Denationalizing science: The contexts of international scientific practice*. Springer Netherlands, pp 233–62.*

Krige J (ed) (1996) *History of CERN, III*. North-Holland.*

Krige J (2001a) Distrust and discovery: The case of the heavy bosons at CERN. *Isis* **92** (3): 517–40.*

Krige J (2001b) Felix Bloch and the creation of a "scientific spirit" at CERN. *Historical Studies in the Physical and Biological Sciences* **32** (1): 57–69.*

Krige J (2001c) The 1984 Nobel Physics Prize for Heterogeneous Engineering. *Minerva* **39** (4): 425–43.*

Krige J (2002) The birth of EMBO and the difficult road to EMBL. *Studies in History and Philosophy of Science Part C: Studies in History and Philosophy of Biological and Biomedical Sciences* **33** (3): 547–64.*

Krige J (2003) The politics of European scientific collaboration. In J Krige and D Pestre (eds) *Companion to science in the twentieth century*. Routledge, pp 897–918.

Krige J (2005) Isidor I. Rabi and CERN. *Physics in Perspective* **7** (2): 150–64.*

Krige J (2008) The peaceful atom as political weapon: Euratom and American foreign policy in the late 1950s. *Historical Studies in the Natural Sciences* **38** (1): 5–44.*

Krige J, A Russo and L Sebesta (2000) *SP-1235: A history of the European Space Agency, 1958–1987, Vol. 2: The story of ESA, 1973 to 1987*. ESA.*

Krücken G, G S Drori and J W Meyer (eds) (2010) *World society: The writings of John W. Meyer*. Oxford University Press.

Latour B and S Woolgar (1986) *Laboratory life: The construction of scientific facts*. Princeton University Press.
Lauto G and F Valentin (2013) How large-scale research facilities connect to global research. *Review of Policy Research* **30** (4): 381–408.*
Liljefors M (2013) Believing in the ESS. In T Kaiserfeld and T O'Dell (eds) *Legitimizing ESS: Big Science as a collaboration across boundaries*. Nordic Academic Press, pp 187–203.*
Lindstrøm M D and K Kropp (2017) Understanding the infrastructure of European Research Infrastructures: The case of the European Social Survey (ESS-ERIC). *Science and Public Policy* **44** (6): 855–64.*
Linné T (2013) The ESS in the local news media. In T Kaiserfeld and T O'Dell (eds) *Legitimizing ESS: Big Science as a collaboration across boundaries*. Nordic Academic Press, pp 85–104.*
Lohrmann E and P Söding (2009) *Von schnellen Teilchen und hellem Licht: 50 Jahre Deutsches Elektronen-Synchrotron DESY*. Wiley-VCH.*
Lozano S, X-P Rodríguez and A Arenas (2014) Atapuerca: evolution of scientific collaboration in an emergent large-scale research infrastructure. *Scientometrics* **98** (2): 1505–20.*
Madsen C (2012) *The jewel on the mountaintop: Fifty years of European Southern Observatory*. Wiley-VCH.*
Marburger J H (2014) The superconducting supercollider and US science policy. *Physics in Perspective* **16** (2): 218–49.*
Martin B R and J Irvine (1981) Internal criteria for scientific choice: An evaluation of research in high-energy physics using electron accelerators. *Minerva* **19** (3): 408–32.*
Martin B R and J Irvine (1984a) CERN: Past performance and future prospects: I. CERN's position in world high-energy physics. *Research Policy* **13** (4): 183–210.*
Martin B R and J Irvine (1984b) CERN: Past performance and future prospects: III. CERN and the future of world high-energy. *Research Policy* **13** (6): 311–42.*
Maxson Jones K, R A Ankeny and R Cook-Deegan (2018) The Bermuda Triangle: The pragmatics, policies, and principles for data sharing in the history of the Human Genome Project. *Journal of the History of Biology* **51** (4): 693–805.*
McLauchlan G and G Hooks (1995) Last of the dinosaurs? Big weapons, Big Science, and the American state from Hiroshima to the end of the Cold War. *Sociological Quarterly* **36** (4): 749–76.*
Meusel E J (1990) Einrichtungen der Großforschung und Wissenstransfer. In H J Schuster (ed) *Handbuch des Wissenschaftstransfers*. Springer, pp 359–71.*
Moed H F, W Glänzel and U Schmoch (2004) Editors' introduction. In H F Moed, W Glänzel and U Schmoch (eds) *Handbook of quantitative science and technology research: The use of publication and patent statistics in studies of S&T systems*. Kluwer Academic Publishers, pp 1–17.
Mongeon P and A Paul-Hus (2016) The journal coverage of Web of Science and Scopus: A comparative analysis. *Scientometrics* **106** (1): 213–28.
Nedeva M (2013) Between the global and the national: Organising European science. *Research Policy* **42** (1): 220–30.
Needell A A (1983) Nuclear reactors and the founding of Brookhaven National Laboratory. *Historical Studies in the Physical Sciences* **14** (1): 93–122.*
Needell A A (1992) From military research to Big Science: Lloyd Berkner and science-statesmanship in the postwar era. In P Galison and B W Hevly (eds)

Big Science: The growth of large-scale research. Stanford University Press, pp 290–311.*
Neumann I B (2012) *At home with the diplomats: Inside a European foreign ministry*. Cornell University Press.
O'Dell T (2013) Mobile spaces of affect. In T Kaiserfeld and T O'Dell (eds) *Legitimizing ESS: Big Science as a collaboration across boundaries*. Nordic Academic Press, pp 67–83.*
Olander B (2013) Social media and research practices in Big Science. In T Kaiserfeld and T O'Dell (eds) *Legitimizing ESS: Big Science as a collaboration across boundaries*. Nordic Academic Press, pp 143–61.*
Panofsky W (1992) SLAC and Big Science: Stanford University. In P Galison and B W Hevly (eds) *Big Science: The growth of large-scale research*. Stanford University Press, pp 129–46.*
Panofsky W and J M Deken (2007) *Panofsky on physics, politics, and peace: Pief remembers*. Springer.*
Papon P (2004) European scientific cooperation and research infrastructures: Past tendencies and future prospects. *Minerva* **42** (1): 61–76.*
Paris E (2001) Lords of the ring: The fight to build the first US electron-positron collider. *Historical Studies in the Physical and Biological Sciences* **31** (2): 355–80.*
Pestre D and J Krige (1992) Some thoughts on the early history of CERN. In P Galison and B W Hevly (eds) *Big Science: The growth of large-scale research*. Stanford University Press, pp 78–99.*
Ploeger J S (2002) The art of science at Fermi National Accelerator Laboratory: The rhetoric of aesthetics and humanism in the national laboratory system in the late 1960s. *History and Technology* **18** (1): 23–49.*
Price D J dS (1963) *Little Science, Big Science*. Columbia University Press.*
Pritzlaff T and F Nullmeier (2011) Capturing practice. *Evidence and Policy: A Journal of Research, Debate and Practice* **7** (2): 137–54.
Qiao L, R Mu and K Chen (2016) Scientific effects of large research infrastructures in China. *Technological Forecasting and Social Change* **112**: 102–12.*
Riordan M (2000) The demise of the superconducting super collider. *Physics in Perspective* **2** (4): 411–25.*
Riordan M (2001) A tale of two cultures: Building the superconducting super collider, 1988–1993. *Historical Studies in the Physical and Biological Sciences* **32** (1): 125–44.*
Rose, F de (2014) *Un diplomate dans le siècle: Souvenirs et anecdotes*. Éditions de Fallois.*
Rosenthal G (2018) *Interpretive social research: An introduction*. Göttingen University Press.
Roy A (2012) Science and the Large Hadron Collider: A probe into instrumentation, periodization and classification. *Dialectical Anthropology* **36** (3–4): 291–316.*
Russo A (2011) Europe's path to Mars: The European Space Agency's Mars Express Mission. *Historical Studies in the Natural Sciences* **41** (2): 123–78.*
Sandell K (2013) Designing for the future. In T Kaiserfeld and T O'Dell (eds) *Legitimizing ESS: Big Science as a collaboration across boundaries*. Nordic Academic Press, pp 163–86.*
Schopper H (2009) *LEP: The lord of the collider rings at CERN 1980–2000: The making, operation and legacy of the world's largest scientific instrument*. Springer.*

Schweber S S (1992) Big Science in context: Cornell and MIT. In P Galison and B W Hevly (eds) *Big Science: The growth of large-scale research.* Stanford University Press, pp 149–83.*
Scott J (2000) *Social network analysis: A handbook*, 2nd ed. SAGE.
Seidel R W (1983) Accelerating science: The postwar transformation of the Lawrence Radiation Laboratory. *Historical Studies in the Physical Sciences* **13** (2): 375–400.*
Seidel R W (1986) A home for Big Science: The Atomic Energy Commission's laboratory system. *Historical Studies in the Physical and Biological Sciences* **16** (1): 135–75.*
Seidel R W (1992) The origins of the Lawrence Berkeley Laboratory. In P Galison and B W Hevly (eds) *Big Science: The growth of large-scale research.* Stanford University Press, pp 21–45.*
Seidel R W (1999) The golden jubilees of Lawrence Berkeley and Los Alamos National Laboratories. *Osiris* **14**: 187–202.*
Seidel R W (2001) The national laboratories of the Atomic Energy Commission in the early Cold War. *Historical Studies in the Physical and Biological Sciences* **32** (1): 145–62.*
Silva F S V, P A Schulz and E C M Noyons (2019) Co-authorship networks and research impact in large research facilities: Benchmarking internal reports and bibliometric databases. *Scientometrics* **118** (1): 93–108.*
Simmonds P, E Kraemer-Mbula, A Horvath, J Stroyan and F Zuijdam (2013) *Big Science and Innovation.* Technopolis.
Simoulin V (2017) An instrument can hide many others: Or how multiple instruments grow into a polymorphic instrumentation. *Social Science Information* **56** (3): 416–33.*
Smith, C L (2007) How the LHC came to be. *Nature* **448**: 281–4.*
Smith R W (1992) The biggest kind of Big Science: Astronomers and the space telescope. In P Galison and B W Hevly (eds) *Big Science: The growth of large-scale research.* Stanford University Press, pp 184–211.*
Stenborg E and M Klintman (2012) Organized local resistance: Investigating a local environmental movement's activities against the ESS. In O Hallonsten (ed) *In pursuit of a promise: Perspectives on the political process to establish the European Spallation Source (ESS) in Lund, Sweden.* Arkiv Academic Press, pp 173–92.*
Strasser B (2003) The transformation of the biological sciences in post-war Europe: EMBO and the early days of European molecular biology research. *EMBO Reports* **4** (6): 540–3.
Sutton J R (1984) Organizational autonomy and professional norms in science: A case study of the Lawrence Livermore Laboratory. *Social Studies of Science* **14** (2): 197–224.*
Teich A H and W H Lambright (1976) The redirection of a large national laboratory. *Minerva* **14** (4): 447–74.*
Tindemans P (2010) Politics of major facilities. *Neutron News* **21** (1): 51–3.*
Traweek S (1988) *Beamtimes and lifetimes: The world of high energy physicists.* Harvard University Press.*
Traweek S (1992) Big Science and colonialist discourse: Building high-energy physics in Japan. In P Galison and B W Hevly (eds) *Big Science: The growth of large-scale research.* Stanford University Press, pp 100–28.*
van der Kruit P (2006) *Five-and-a-half years in ESO Council.* Available at www.astro.rug.nl/~vdkruit/jea3/homepage/president.pdf (last accessed December 2, 2019).*

Vermeulen N (2016) Big biology: Supersizing science during the emergence of the 21st century. *NTM Zeitschrift für Geschichte der Wissenschaften, Technik und Medizin* **24** (2): 195–223.*

von Platen S (2013) Reaching the inside from the outside? In T Kaiserfeld and T O'Dell (eds) *Legitimizing ESS: Big Science as a collaboration across boundaries*. Nordic Academic Press, pp 123–42.*

Wang Z (1995) The politics of Big Science in the Cold War: PSAC and the funding of SLAC. *Historical Studies in the Physical and Biological Sciences* **25** (2): 329–56.*

Wasserman S and K Faust (1994) *Social network analysis: Methods and applications*. Cambridge University Press.

Weinberg A M (1962) The federal laboratories and science education: By playing a greater role in education, Big Science can diminish the manpower shortage it has created. *Science* **136** (3510): 27–30.

Wellock T R (2012) Engineering uncertainty and bureaucratic crisis at the Atomic Energy Commission, 1964–1973. *Technology and Culture* **53** (4): 846–84.*

Westfall C (1989) Fermilab: Founding the first US "truly national laboratory". In F A J L James (ed) *The development of the laboratory*. Palgrave Macmillan, pp 184–217.*

Westfall C (2002) A tale of two more laboratories: Readying for research at Fermilab and Jefferson Laboratory. *Historical Studies in the Physical and Biological Sciences* **32** (2): 369–407.*

Westfall C (2003) Rethinking Big Science: Modest, Mezzo, grand science and the development of the Bevalac, 1971–1993. *Isis* **94** (1): 30–56.*

Westfall C (2008) Retooling for the future: Launching the Advanced Light Source at Lawrence's Laboratory, 1980–1986. *Historical Studies in the Natural Sciences* **38** (4): 569–609.*

Westfall C (2010) Surviving to tell the tale: Argonne's Intense Pulsed Neutron Source from an ecosystem perspective. *Historical Studies in the Natural Sciences* **40** (3): 350–98.*

Westfall C (2012) Institutional persistence and the material transformation of the US national labs: The curious story of the advent of the Advanced Photon Source. *Science and Public Policy* **39** (4): 439–49.*

Westfall C (2017) Between the lines: A first-person account of Berkeley's loss of Fermilab. *Physics in Perspective* **19** (2): 91–104.*

Westfall C and L Hoddeson (1996) Thinking small in Big Science: The founding of Fermilab, 1960–1972. *Technology and Culture* **37** (3): 457–92.*

Westwick P J (2000) Secret science: A classified community in the national laboratories. *Minerva* **38** (4): 363–91.*

Westwick P J (2007) Reengineering engineers: Management philosophies at JPL in the 1990s. *Technology and Culture* **48** (1): 67–91.*

Widmalm S (1993) Big science in a small country: Sweden and CERN II. In S Lindqvist (ed) *Center on the periphery: Historical aspects of 20th-century Swedish physics*. Science History Publications, pp 107–40.*

Wilson G and C G Herndl (2007) Boundary objects as rhetorical exigence: Knowledge mapping and interdisciplinary cooperation at the Los Alamos National Laboratory. *Journal of Business and Technical Communication* **21** (2): 129–54.*

Wojcicki S (2008) The supercollider: The pre-Texas days: A personal recollection of its birth and Berkeley years. *Reviews of Accelerator Science and Technology* **1** (1): 259–302.*

Wojcicki S (2009) The supercollider: The Texas days: A personal recollection of its short life and demise. *Reviews of Accelerator Science and Technology* **2** (1): 265–301.*

Yu H, J B Wested and T Minssen (2017) Innovation and intellectual property policies in European Research Infrastructure Consortia, Part I: The case of the European Spallation Source ERIC. *Journal of Intellectual Property Law and Practice* **12** (5): 384–97.*

Zhang J, M S Vogeley and C Chen (2011) Scientometrics of big science: A case study of research in the Sloan Digital Sky Survey. *Scientometrics* **86** (1): 1–14.*

3. The role of European Big Science in the (geo)political challenges of the twentieth and twenty-first centuries

Katharina C. Cramer

1. INTRODUCTION

Change is a topical issue of recent scholarly research on Big Science and Research Infrastructures (RIs). This has not only been highlighted in the introductory chapter of this book and runs as a red thread through most of the individual contributions, but it is also covered by current research attempts such as work on New Big Science (Rekers and Sandell 2016; Crease and Westfall 2016) or Big Science Transformed (Hallonsten 2016). While Big Science in the early Cold War years, most notably in Europe and the United States (US), often inscribed into political logics and rhetoric that saw the pre-eminence in science and technology as a crucial condition for national security and power (Seidel 1992; Crease 1999; Hoddeson et al 2008), this started to change, most importantly, since the 1970s and 1980s as political framework conditions in and geopolitical contours of the Western world changed (Hallonsten and Heinze 2012, 2013; Westfall 2012). This development also links to a stepwise reorientation of science policy agendas due to a general shift of the role of science and knowledge as the fundament for modern societies and economies (Hallonsten 2016; Cramer et al, ch 1 in this volume). A nuanced analysis of these and similar variations of change certainly adds to the large body of different scholarly investigations of the history, politics or organization of the two categories Big Science and RIs. It also improves our understanding of how and to what extent these two categories relate to each other.

However, when it comes to change, continuity also has its role to play. While, as described above and in the introductory chapter of this book, socio-political contexts changed profoundly throughout the most recent decades and with them the purpose and *raison d'être* of many of Europe's Big Science projects, at the same time, the way these projects were made in reality did not change much. On the one hand, since the early Cold War years, these projects in Europe still come into being through intergovernmental ad hoc agreements. The dynamics and politics of negotiating, bargaining and deal making among and between governmental representatives still remain at the core of these projects. This observation does not only call, in the context of this chapter, for the continued use of Big Science instead of big science, the former in capital letters "as a rhetorical construction" pointing to the particular dynamics of large-scale research following the end of World War II (Capshew and Rader 1992: 22; see also Cramer et al, ch 1 in this volume). It also points to a more general dynamic, namely that Big Science projects are European politics by other means, and that therefore, the historical developments of these projects often stand as proxies for broader political controversies and diplomatic struggles (Krige 2003, 2006; Hallonsten 2014, 2016). Put differently, the rationales of collaborative Big Science projects in Europe escalate beyond symbolic and prestigious representations through their size and costs. But to the extent that politics matter, Big Science projects possess a structural function as well as a role in forging relationships between countries.

Connecting to this, the present chapter shows that intergovernmental negotiations as well as multilateral compromise and controversy during the founding phases of several Big Science projects link to periods of political disarray and diplomatic struggle and that, more generally, Big Science collaborations have a significant role to play in times of political crisis in Europe. To this end, this chapter investigates the founding histories of several collaborative Big Science projects in the natural sciences that were established throughout the last decades in (Western) Europe by a varying number of countries. The projects examined are the European Organization for Nuclear Research (CERN), the European Southern Observatory (ESO), the European Space Research Organisation (ESRO) together with the European Space Vehicle Launcher Development (ELDO), the Institut Laue-Langevin (ILL), the European Synchrotron Radiation Facility (ESRF), the European Transonic Wind Tunnel (ETW), the European X-Ray Free-Electron Laser (European XFEL) and the Facility for Antiproton and Ion Research (FAIR). While most of these projects are single-sited, ESO is an exception here with its observatory based in Chile and the administrative and organizational headquarters in Germany. The scientific purposes of the different projects are diverse and range from fundamental research in particle physics (in the case of CERN) to multidisciplinary research in the fields of molecular biology, chemistry or life sciences which coexists with physics, chemistry, materials science and engineering at the European XFEL and ESRF, among others.

This chapter contextualizes the political histories of the projects mentioned above and connects them to dynamics of policymaking and political agenda setting in Europe as well as the European Economic Community (EEC) and the European Union (EU). In the following, the EEC refers to the community of Europe pre-1992 (when the Maastricht Treaty was signed) and the EU refers to the same collaborative thereafter. EEC/EU and Europe are not used synonymously, but they sometimes appear side to side to emphasize that the formation and the historical development of Europe should not be conflated with the EEC/EU and/or the European integration process.

The chapter is based on previous scholarly research on the history, politics and organization of Big Science in Europe (e.g. Krige 2003, 2006; Hallonsten 2014, 2016; Hallonsten and Heinze 2012, 2013) as well as major works on the history and politics of Europe and the EEC/EU (e.g. Middlemas 1995; Judt 2005; Patel 2018). The investigations of the individual projects benefited much from case-related historical investigations (e.g. Blaauw 1991, 1997; Cramer 2017, 2020; Herman et al 1987; Krige et al 2000; Schmied 1990a, 1990b).

The chapter is structured as follows. Section 2 focuses on the different ways of how politics matter during the founding phases of European Big Science projects and to what extent the politics of these projects formally and informally link to common European politics and policy activities. Section 3 introduces the role of European Big Science in times of political crisis. The main argument is that although Big Science projects in Europe remain based on intergovernmental agreement and to a large extent formally disentangled from common and formal European policymaking, the establishment of these Big Science projects constitutes additional forums for European countries to find common political ground to jointly overcome controversy, conflict and disarray. Then follows a review, on a case-by-case basis, of the founding histories of several Big Science projects in Europe and an exploration of their roles in politically indeterminate situations. The chapter ends with a concluding section that carefully draws the main findings together and summarizes the role of European Big Science in the (geo)political challenges of the late twentieth and twenty-first centuries.

2. POLITICS AND POLICY OF BIG SCIENCE IN EUROPE

The politics of Big Science in Europe go beyond the traditional goals of science diplomacy and similar concepts that promote scientific collaboration for the sake of peace and friendly international relations (e.g. Moedas 2016). As this section highlights, patterns of diplomatic and political relations among countries in and around Europe, and moments of deepening European integration or times of European political crisis and upheavals, are mirrored by the politics played out during the establishment, construction and operation of collaborative Big Science projects. The reasons and motivations to collaborate and to reach intergovernmental agreement among several countries in Europe since the mid-twentieth century were often framed by political boundary conditions. In this regard, collaborative Big Science in Europe was and remains not much an issue of shared intrinsic motivation and joining efforts for the common good. Rather, the creation of Big Science projects was (and remains) to a great extent dictated and impacted by national political interests and the dynamics of high-level diplomacy (Krige 2003, 2006; Hallonsten 2012, 2014, 2016).

Politics is here defined and understood as situations of negotiation, bargaining and package-deal creation among and between countries represented by administrators or governmental representatives. It can, moreover, be interpreted as a way of containing the power of the other partner, framing political relationships, defining space and territory, as well as the pursuit of national interests and strategies. This approach relates to similar perspectives and understandings in sociology and history that regard science and politics as "mutual resource systems" (Trischler 2010). These are characterized as distinct but interrelated and/or complementary realms that target the interfaces between science and politics to investigate and understand their historical developments and dynamics (e.g. Weingart 1999; Ash 2010; Trischler 2010).

Collaborative Big Science projects in Europe are based on intergovernmental agreements that come into being through ad hoc agreements signed by a varying number of countries. The importance of intergovernmental agreement and compromise escalates beyond the setting of boundary conditions for scientific research. Among other things, intergovernmental frameworks offer the possibility for the individual collaborating countries to bring national agendas and priorities to the table. Moreover, these settings often provide an impetus for countries to find common (political) ground beyond science. Situations of negotiating and bargaining also fulfill the function of circumventing diplomatic taboos and of demonstrating a kind of unity and collective identity in which the national, the multilateral, the political and the scientific contexts overlap. Since the mid-twentieth century these intergovernmental settings of Big Science projects in Europe thus are nothing less than European politics by other means (e.g. Krige 2003, 2006; Hallonsten 2014, 2016; Cramer 2020).

Although there is much to suggest that the politics of Big Science bear significant relevance to European politics and policymaking, throughout the second half of the twentieth century these projects neither constituted an official body of the EEC/EU, nor were they formally entangled with European integration mechanisms (Cramer 2020). It was only by the early 2000s that the operation, coordination and management of collaborative Big Science projects acquired a visible and prominent place on the political agenda of the EU – under the umbrella term of Research Infrastructures.

In the late 1980s and the 1990s, the common funding instruments of the EEC/EU, the Framework Programmes (FPs) for Research and Technological

Development, devoted attention to the issue of Big Science namely through measures that should facilitate transnational access to existing facilities and that should enable researchers to improve their scientific performance and outreach. Yet, this was only a minor aspect of these early FPs and the budget was small compared to the other major initiatives of the FPs at this time (Hallonsten 2020; Stahlecker and Kroll 2013).

Since the early 2000s, the EU has developed a supportive framework to better coordinate and manage collaborative Big Science projects in Europe. The most visible manifestation of how Big Science became a topical issue for the EU probably is the politically induced introduction and frequent use of the concept of RI by the EU Commission in many of its policy documents and strategic reports (Hallonsten 2020). Moreover, the European Strategy Forum on Research Infrastructures (ESFRI), that emerged from a former informal round table of high political delegates from the science, technology and/or research ministries of the member countries, was founded in 2002. The publication of the first ESFRI roadmap in 2006 intended to provide a systematic and coordinated approach to plan and prioritize large-scale research projects in Europe (Bolliger and Griffiths, ch 5 in this volume). In 2008, the new organizational form of the European Research Infrastructure Consortium (ERIC) was created through a specific legal framework, and aimed to simplify collaborative research projects and facilitate the set-up of new RIs in Europe (Moskovko et al 2019; Moskovko, ch 6 in this volume).

Summarizing these recent developments, it seems reasonable to argue that in the last decades the EU has experienced the stepwise formation of an RI policy. There now exists a strategic link between Big Science and RIs to the extent that the EU Commission has established and implemented several policy tools (ESFRI, ERIC) that actively consider, under the common label of RIs, several (but not all) of the already existing and/or planned collaborative and intergovernmental Big Science projects. While several of these projects were included in one or several of ESFRI's roadmaps such as the European XFEL or the upgrade programs of ESRF and ILL, several of them did not, such as ESO or CERN. While several projects also adopted the ERIC framework, such as the European Spallation Source (ESS), several of them, such as CERN, ESRF or European XFEL, did not. They still operate with different organizational frameworks, such as limited liability companies (in the case of ESRF and European XFEL) or as intergovernmental treaty organizations (in the case of CERN and ESO).

If these considerations are any guide then it is that the emerging RI policy constitutes a potentially new and somehow alternative dimension of how to organize, manage and establish particularly large research projects in Europe (Ulnicane, ch 4 in this volume). But these current policy efforts of the EU and its political interest in RIs did not (yet) transform the basic principles of

Big Science in Europe, namely that intergovernmental agreements by state groups of varying size, negotiated among ministerial and governmental representatives, remain (at least) the widespread modus operandi of collaborative large-scale research projects in Europe.

3. BIG SCIENCE IN TIMES OF POLITICAL CRISIS

From the crisis around the Constitutional Treaty in 2005, the financial crisis in 2008, to "Brexit" as the potential disintegration of the United Kingdom from the EU, scholarly research in recent years covering these and similar situations suggests that crisis is the "new normal" of the EU in the twenty-first century (Cabane and Lodge 2018). Changing international contexts and emerging global challenges, such as the end of the Cold War, the powerful political and economic rise of China and Russia, or accelerating globalization tendencies, certainly constitute additional aspects of the widely shared public perception of Europe's current critical state and the many different (geo)political challenges it has had to face in the past decades (e.g. Habermas 2012; Dinan et al 2017; Cross 2018; Demetriou 2015; Krzyżanowski 2019; van Middelaar 2019). More generally, scholarly research on European integration also frequently refers to the history of the EEC/EU as a history of crises in the sense that "the notion of crisis is associated with either the definitive failure of the process of cooperation in Europe, or its relaunch" (Warlouzet 2014: 98).

What all these situations share is the confrontation with something new and unprecedented that may fundamentally threaten or alter existing structures or contexts (e.g. Koselleck and Richter 2006; Runciman 2016). These situations call for political decisions and actions to be taken in order for Europe and the EEC/EU to (re)affirm its standing as a political entity, economic union or shared cultural space. Yet, moments of crisis, uncertainty and disarray are not only phenomena of Europe's recent past but indeed include very long historical trajectories. Although the creation of the EEC in 1957, together with its sibling Euratom, often carried the promise of a peace project, this did not mean that critical periods and moments of (political) disarray remained absent.

On the one hand, the EEC/EU per se did not bring peace and political stability to the continent, but rather its various institutions, processes and principles that are testament to the political integration process and that created forums and shared spaces to find common political ground to jointly overcome controversy, conflict and disarray (Patel 2018). On the other hand it is, however, not a *conditio sine qua non* that policy fields have to be formally and politically integrated into EEC/EU's framework in order to contribute to challenging political disarray. Rather, to the extent that Big Science projects in Europe were (and remain) formally disentangled from common European policy mechanisms and integration attempts, in politically indeterminate periods, the intergovernmental settings to establish Big Science projects nevertheless contributed to the creation of a sense of belonging among several European countries and provided an impetus for political consolidation beyond the realm of scientific collaboration.

Reconciling from above, it is one of the core messages of this chapter that the political histories of Big Science projects in Europe and their direct connections to European politics and policymaking offer additional insights into how the EU and its individual member countries face and challenge political disarray. In this regard, the next sections engage with the founding histories of several collaborative and intergovernmental Big Science projects in Europe that were established throughout the second half of the twentieth century and early twenty-first century to explore how and to what extent the (political) histories of these projects link to moments of political disarray and controversy.

4. CASE STUDIES

4.1 CERN and ESO

The political situation in Europe in the late 1940s and 1950s was fragile and indeterminate. But European countries became more and more aware that common challenges could better be solved in an intergovernmental rather than a national context, and thus also provide containment of the other's increasing power on the European continent and contribute to the realization of national ambitions (Middlemas 1995; Judt 2005). To this end, this period saw the creation of many different intergovernmental organizations in Europe. Next to the well-known predecessors of today's EU, the European Coal and Steel Community in 1952 and Euratom and the EEC in 1957, these included less known organizations such as the Council of Europe and the Western European Union. In one regard or the other, the implementation of these and similar organizations intended to create a sustained framework for cooperation and integration among European countries in order to lay a fundament for long-term political stability (Patel 2018: 23–5). The creation of CERN in 1954 and ESO in 1962 adds to this picture in the sense that these projects, next to their scientific ambitions, can also be considered efforts of sustained political cooperation among several countries in Europe during the politically indeterminate postwar years.

CERN, located at the French-Swiss border close to Geneva in Switzerland, was established in 1954 by 12 countries: Belgium, Denmark, France, Greece, Italy, the Netherlands, Norway, Sweden, Switzerland, the United Kingdom, West Germany and Yugoslavia. The project's core mission was to carry out fundamental research in the field of nuclear physics, and it remains today the leading research center in Europe for fundamental research in particle physics.

First proposed in the mid-1940s by influential scientists from Europe and the US, the eventual creation of CERN and its location on politically neutral ground in Switzerland was much welcomed and supported by the US to strengthen regional collaboration and Western integration in Europe, among other things (Hermann et al 1987; Krige 2006).

Promoted as a collaborative peace project, CERN became Europe's first experience in large-scale research collaboration. The straightforward creation of CERN, compared to many later Big Science projects that experienced serious time overruns (see Hallonsten, ch 10 in this volume), benefited much from the monetary generosity of the larger European countries in this period and the high significance of atomic energy, and related to this, nuclear research and technology for postwar economic prosperity. The creation of CERN was not only a milestone in the process of achieving political unity in (Western) Europe. Considering the politically and diplomatically fragile postwar years, CERN also became a role model for Europe's own long-term competitiveness strategy that would help the continent to reestablish a leading role in global science and politics (Krige 2003, 2006; Hallonsten 2014).

The core mission of ESO is to establish and maintain facilities for ground-based astronomy as well as to manage access to these facilities. While the astronomical equipment is based in northern Chile, the headquarters is in Garching, close to Munich, Germany. Similar to CERN, the initial idea to create ESO arose in the early 1950s among a group of European and US-American influential scientists, and it was also inscribed into a similar socio-political context of the postwar years in Europe. But its eventual creation was delayed several years due to political constraints among several of its potential member countries. It was eventually founded in 1962 by Belgium, France, the Netherlands, Sweden and West Germany (Blaauw 1991, 1997; Woltjer 2006; Cesarsky and Madsen 2007).

The realization of ESO was, however, not as easy as it was for CERN, although its organizational and legal frameworks largely followed those of CERN. Yet, ground-based astronomy did not have the same prestige attached to it as nuclear physics. Moreover, the observatory needed the clear sky of the southern hemisphere and hence no potential European member country could act as host (Blaauw 1991; Hallonsten 2012, 2014). Generally speaking, the selection of a site is what usually, in a literal sense, puts a European Big Science project in place. Harsh negotiations over location often lead to the agreement on a site premium, to be paid by the hosting country that should compensate the many economic and symbolic benefits by having the facility built on its own territory. In this regard, site selection seldom remains a (simple) question of geography but is often considered a crucial pre-condition for any other preparatory activity, such as negotiations over financial shares or legal and organizational frameworks (Hallonsten 2014). The ESO project lacked this competitive process of site selection among its potential member countries, because it was clear that the observatory could not be located on the European continent. Although the project was backed by leading scientists and the promise of new and fundamental scientific discoveries, it thus became a challenge to motivate several governments to commit to a research facility that would, by definition, never be built on their own territory. The United Kingdom even withdrew from the project in 1960 because it preferred to invest in an astronomical large-scale project located in Australia, as part of the British Commonwealth (Blaauw 1997: 111; Krige 2003: 906).

The eventual realization of ESO was enabled by, on the one hand, a grant from the US-American Ford Foundation under the condition that at least four out of five potential member countries would sign the convention, and on the other hand, decisive contributions from West Germany and France, who paid one third each of the preparation costs of the project during its founding phase. The remaining third was shared by Belgium, the Netherlands and Sweden. German contributions rose to 49 percent of the total preparation costs in 1958 and 1959 when France had to temporarily withdraw all contributions due to internal political upheavals (Blaauw 1991: 232ff).

The bilateral alliance between France and West Germany was a key feature of ESO's founding phase, and also drove and shaped the course of events during ESO's early history. The close cooperation of these two countries is an early testament of French–German reconciliation after the end of World War II and highlights that Big Science projects, such as ESO, may represent "a particularly useful first and tentative step in a politically delicate context of alliance building", as historian of science John Krige puts it (2003: 904). In this context, it is hence worth highlighting that ESO's formal creation in 1962 was paralleled by the preparation for the Elysée Treaty that was signed one year later by French president Charles de Gaulle and German chancellor Konrad Adenauer. This treaty is generally regarded as a major bilateral effort on political, scientific, cultural and military cooperation between the two countries (Defrance and Pfeil 2013; Defrance 2016). The creation of ESO, backed by a strong French–German partnership, certainly benefited from this political atmosphere of reviving bilateral relations.

4.2 ESRO and ELDO

ESRO and ELDO were created in the early 1960s, a few years after the so-called "Sputnik shock" of 1957, when the Soviet Union launched the first satellite Sputnik into orbit. This was a decisive event, after which competition in space between the Soviet Union and the US escalated: The US-American National Aeronautics and Space Administration (NASA) was created in 1958. In 1961, the Soviet cosmonaut Yuri Gagarin was the first human in outer space. In 1969 the US-American Apollo 11 landed as the first manned mission on the moon. For both the US and the Soviet Union, space science and policy became key means of demonstrating national power as well as scientific and technological superiority (Krige et al 2000: 5ff).

In this highly political and tension-laden context, several European countries including France, Italy, the United Kingdom and West Germany had begun to develop national space programs. But they also decided to join forces to create a European approach through ESRO and ELDO. ESRO was established in 1962 by ten European countries: Belgium, Denmark, France, Italy, the Netherlands, Spain, Sweden, Switzerland, the United Kingdom and West Germany. ESRO's main mission was to conduct space-related research and to develop satellites. ELDO was founded the same year by the same group of countries to build the necessary launcher technology. There were several reasons for keeping these two projects, science-oriented ESRO and technology-oriented ELDO, formally apart (see below). In the mid-1970s, the two projects were merged into today's European Space Agency (Krige et al 2000).

Certainly, both projects served as a political instrument to counter the US-American and Soviet space activities. Although ELDO's launcher program did not prove successful in the end, the technological ambitions that stood behind it can be regarded as a means to retain autonomy by not being dependent on launch equipment from other countries (Krige et al 2000: 9ff; Krige 1992). But the aim of both projects was also to do "more than simply compete with the superpowers in space" (Kariya 2009: 35). First, ESRO was conceived outside of military interest and while military and security state-level interests mattered in the case of ELDO, both projects were planned in conjunction with the industrial policies of their member states to protect national activities (Krige 2003: 907). This certainly cohered with the socio-political context of the 1960s in Europe, that "experienced not only a quantitative growth in the interest in space science and technology, but a progressive, though indecisive, reorientation of European interest away from 'pure' space science toward a kind of activity linked not only to military but also to commercial interests, especially in the field of satellites" (Krige et al 2000: 387).

ESRO and ELDO's founding rationales escalated beyond the mere symbolism and prestige of being part of a politically driven space race. But they closely linked to commercial considerations that can hence be interpreted as a key aspect that made European countries commit to this project. Beyond that, there is also much to suggest that ESRO and ELDO were initiated with the vision of creating an organization with a specific European character (Kariya 2009; Krige et al 2000). Put differently, "[s]cience had always possessed an international character, but under ESRO it was specifically European. As a standalone organization, its potential to integrate Europe was limited, but it added a layer to the history of European experiences. Its convocation was an important step in the process of constructing a collective identity" (Kariya 2009: 32).

In this context, the comprehensive European character did not only refer to the (most) powerful European countries, but also included smaller as well as neutral countries in Europe. Two aspects are indicative in this regard. First, the United Kingdom and France were the strongest European countries in space science and research at that time and they could have easily created a bilateral project. But "both countries consciously sacrificed efficiency, comparative gains, and freely permitted the transfer of costly technology in order to form a European organization" (Kariya 2009: 36). Second, one reason to keep ESRO and ELDO formally apart was that membership would have become quite expensive: ELDO required heavy financial investments to develop the launcher technology, whereas ESRO with its focus on project-based research was expected to be a less costly endeavor. This was particularly important for the smaller European countries that could not afford costly membership in both organizations. Another reason was that a combination of both programs would have meant excluding the (politically) neutral countries such as Norway or Switzerland from membership that did not want to finance potentially military-oriented technology which could have been the case for ELDO and the launcher technology (Krige et al 2000: 37; Kariya 2009: 33).

This rationale of a specific European character can thus be understood as an attempt to define what Europe is and what role it could play in this politically delicate period in the early Cold War. Participation in this project thus became a powerful identificatory moment in the sense that the members of ESRO and ELDO tried to create a sense of belonging among European countries as a way and a means to clearly differentiate itself from the political activities of the US and the Soviet Union and to establish a specific European project.

4.3 ILL

ILL in Grenoble, France was established in 1966 as a French–German bilateral effort. It became the first purpose-built reactor in Europe for producing neutrons for the investigation of the structure and properties of materials. After initial participation, the United Kingdom withdrew from the project because of budget constraints, but also because the government had turned to other priorities, including the establishment of a national research reactor. The French–German tandem took over the task of realizing the project but the United Kingdom (re)joined in 1974 as a third partner (Bacon 1987; Rush 2015; Atkinson 1997; Jacrot 2006).

The founding history of the ILL shows impressive links to European politics at the time, in mainly two regards. First, planning and preparations for the ILL took place during a time of deep political crisis between France and the United Kingdom. More concretely, the British withdrawal from the ILL project in its planning phase and its return as a third member in 1974 mirrors the overall complicated political situation with the British membership application to the EEC in the 1960s. British applications to join the EEC were vetoed by French president de Gaulle in 1963 and 1967. Only after de Gaulle resigned in 1969 and the pro-European British prime minister Edward Heath was elected in 1970, could the British strive towards EEC membership be resolved, and the country joined the EEC in 1973 (Judt 2005: 292, 307f.). One year later in 1974, the United Kingdom officially became a third member of the ILL.

Second, the ILL is also a remarkable example of what France and West Germany were able to achieve in large-scale scientific collaboration in the 1960s and 1970s (Hallonsten 2016: 90; Jacrot 2006; D'Ippolito and Rüling, ch 11 in this volume). As described in the last section, collaboration in Big Science between these two countries can be regarded as an effort to reconcile after World War II. With regard to the fact that the ILL was founded only several years after the signing of the Elysée treaty that can be considered, as mentioned above, as a major effort in French–German relations, historian of science Dominique Pestre argues that the scientists that initiated the ILL project benefited from a "window of opportunity" at this time "offered by the improvement of Franco-German relations initiated by Adenauer and de Gaulle" (Pestre 1997: 138). This successful partnership certainly also paved the way to establish the ESRF and the ETW in the 1980s (see below).

4.4 ESRF and ETW

The early 1980s were a difficult time for the member countries of the EEC and the pursuit of the European integration process. Unsolved intra-European political issues, ranging from budgetary questions to a reconsideration of the common agricultural policy, loomed in parallel to challenges from abroad, such as the growing economic competition from the US and Japan. The late 1970s had brought the second oil crisis and an economic recession, high inflation and unemployment to Europe.

In this gridlocked situation that fundamentally halted further efforts in European political integration, France and West Germany decided to join forces to promote and mediate compromise between the member countries of the EEC and to push national interests forward (Saunier 2001). During the years 1983 and 1984 the two countries "witnessed an intensification of bilateral contacts on all levels" (Krotz and Schild 2013: 119). Moreover, French president Mitterrand had also conducted a "highly active shuttle diplomacy" (Krotz and Schild 2013: 119) at this time, visiting the heads of governments of the other European member states and developing a close personal relationship with each one of them (Saunier 2001: 482). Keith Middlemas supplements this picture, noting that all pending issues and open questions on the European political agenda "had been brought into line, largely by Mitterrand, now at his peak, yet ever conscious of the need for West German backing" (1995: 1079). The outcome of these French–German efforts was a complex and densely meshed net of package deals, by which compromises on common political concerns were concluded by satisfying, in turn, the particular national interests of the various EEC member countries (Krotz and Schild 2013: 119–20).

While the EEC member countries had thus eventually found common political ground, largely due to the strong and active French–German partnership, this political context was also key to set political sails for the ESRF project. Based on an initiative of leading scientists in Europe, the project developed under the auspices of the European Science Foundation in the late 1970s and 1980s (Cramer 2017; Schmied 1990a, 1990b). The science and research ministers of France and West Germany had already started to officially meet frequently throughout the years 1983 and 1984. The (political) breakthrough regarding the ESRF project then came in the fall of 1984 when France and West Germany agreed on initial funding and a site in Grenoble. Although this bilateral decision was contested by other countries, it provided a baseline that was joined by several countries a few years later (Cramer 2017, 2020).

The early history of the ESRF closely links to the ETW project. The ETW is located in Cologne, Germany and was established by four European countries in 1985: France, the Netherlands, the United Kingdom and West Germany. The ETW project was of strong interest for the national industries in the civil and military aviation sectors. It was expected at that time that the competitiveness of the European aviation industries in the future would depend largely on the capabilities and capacities of such simulation and testing facilities (Hirschel et al 2004; van der Bliek 1996).

The French–German agreement in 1984 on a site for the ESRF was part of a broader package deal between the two countries that located the ETW at Cologne. In 1983, the science ministers of France and West Germany had agreed on the creation of the wind tunnel project on a bilateral basis, but they could not take a final decision with regard to the site: West Germany preferred Cologne, and France was in favor of Toulouse, given that the Airbus company was already located there (Cramer 2017).

Summarizing these considerations, the founding histories of both projects closely link to the strong role of the French–German tandem in European politics in the 1980s. The bilateral decision on a site for both the ESRF and the ETW as well as on initial funding for the ESRF was a remarkable French–German demonstration of their political strength and unity in Europe at that time. In this regard, both projects provided an additional (collaborative) impetus for European integration to revive during this critical period.

4.5 European XFEL and FAIR

The founding histories of both European XFEL and FAIR have their roots in the early 1990s when the end of the Cold War, the dissolution of the Soviet Union and the fall of the Berlin Wall had only shortly before swept away political certainties and geopolitical balances that had guided and dominated the continent throughout the Cold War period.

These developments also translated into new forms of political alliances and multilateral settings. The degree of integration of Russia as an important actor into the political agendas of both individual European countries and the EU remains until today a crucial diplomatic concern. While Russian formal membership in Western alliances such as the EU or the North Atlantic Treaty Organization was never seriously put on the agenda, diplomatic relations with Russia gained new weight when the external borders of the EU were pushed closer to Russia with the enlargement rounds in 2004 and 2007 (Timmins and Gower 2009; Kühnhardt 2008; Judt 2005).

Russia's role in the post-Cold War is not only a point on the agenda of European politics and diplomacy, but this concern also translates into the recent history of Big Science in Europe. Certainly, scientific collaboration between East and West was not at all absent during the Cold War. But formal membership for Russia in Big Science projects in (Western) Europe only became an accepted new reality in the post-Cold War period. For instance, in 1999 Russia acquired observer status in CERN. In 2012, the country applied for associate membership but the issue is still pending (Russian Government 2019). In 2014, Russia became a full member of the ESRF, contributing 6 percent of the operation costs. This led to a major redistribution of the financial shares of the other members. Among other projects, Russian membership also particularly matters with regard to the European XFEL and FAIR projects (Cramer 2020).

The European XFEL in Hamburg, Germany is a recent effort for the realization of a state-of-the-art free electron laser based on an intergovernmental agreement that was signed in 2009 by 12 countries: Denmark, France, Germany, Greece, Hungary, Italy, Poland, Russia, Slovakia, Spain, Sweden and Switzerland. FAIR is an accelerator complex for nuclear physics at Darmstadt, connected to the GSI Helmholtz Centre for Heavy Ion Research. An intergovernmental agreement was signed to establish FAIR in 2010 by Finland, France, Germany, India, Poland, Romania, Russia, Slovenia and Sweden.

Russia became the second biggest shareholder in the European XFEL project, contributing €250 million or 23.1 percent of the construction costs, in 2009. Russia also became a member of the FAIR project, in 2010, with a similar substantial contribution of €178 million or 17.3 percent of the construction costs (European XFEL 2009; FAIR 2010). Taken together, Russian financial contributions to these two facilities represent "the first time that Russia is investing such amounts of financing in research facilities which are not located on its own territory" (European Commission 2013). This large financial contribution certainly helped the two projects during their founding phases, at a time when other European countries and potential collaborating partners were quite reluctant to fund such costly efforts. One main concern of these potential partner countries certainly was the outbreak of the financial crisis in 2008/2009 and the serious economic consequences and financial difficulties that some countries such as Spain, Italy and the United Kingdom had to face (Hallonsten 2012).

Russian membership in several (Western) European Big Science projects and its enormous investment in the European XFEL and FAIR projects are symptomatic for a wider and thorough reorientation of Russian science policy and the restructuring of its national science system following the economic and political upheavals in the 1990s. These efforts eventually led to the 2011 Russian Megascience Initiative that intends to establish several large research facilities with international contributions on Russian territory that mirror those in Western Europe scientifically and technologically (Prime Minister of the Russian Federation 2011). While (Western) European Big Science projects seemingly carry strong territorial implications for Russia, the Megascience Initiative can thus be interpreted as a strategy for Russia to demand (Western) European countries for a return of investment on large scientific projects in Russia and, importantly, on Russian territory.

In this regard, there is much to suggest that the emerging role of Russia in (Western) European Big Science in recent decades clearly adds a new geopolitical component to large-scale research in Europe. Increasing Russian participation in large and collaborative scientific projects in (Western) Europe and the substantial Russian contributions to each of these projects can be considered a strategy to get involved in European political affairs, through the means of science and technology, without proposing or negotiating some kind of formal membership in the arrangement of the EU. This opens a Pandora's box of how to define the often complex and indeterminate political and diplomatic relations with Russia in the post-Cold War era; an issue that has often been circumvented by the EU's high political agendas (van Middelaar 2019).

5. CONCLUDING DISCUSSION

A remarkable portfolio of collaborative Big Science projects was established in Europe throughout the second half of the twentieth century and the early twenty-first century. This chapter reviewed and contextualized the founding histories of several of these projects to improve scholarly understanding on how and to what extent the politics of these Big Science projects in Europe mirror and resonate broader political controversies and diplomatic struggles.

Summarizing empirical evidence and the main findings of this chapter, three aspects are worth highlighting that have seemingly run as a red thread through the history of Big Science in Europe since the early postwar years and that refrain the dynamics of European politics in periods of controversy and disarray: (1) the strong role of the French–German partnership; (2) the rocky relationship with the United Kingdom; and (3) the emerging role of Russia in (Western) European Big Science projects since the 1990s.

First, the French–German partnership provided a political baseline of several Big Science projects through joint pledges of necessary initial funding or joint decisions on sites. Next to the role of these two countries in the creation of ESO, the ILL, ESRF or ETW, as described above, they also mattered for the establishment of the European Molecular Biology Laboratory (EMBL) in 1976 and the Institut de Radioastronomie Millimétrique (IRAM) in 1979. To summarize this historical development, the two countries did not only diplomatically reconcile after the end of World War II and become a powerful political entente in the Western European context that was often indicative for the challenge of politically indeterminate situations. They were also major driving forces behind the creation of collaborative Big Science projects; most importantly when the overall political situation in Europe was fragile or during periods of multilateral political disarray.

Second, participation of the United Kingdom in several Big Science projects often proved difficult and controversial. As highlighted above, in the cases of ESO, the ILL and European XFEL, the country withdrew at one point or the other during the founding phase, either because of budget constraints and/or because national interests had turned to other priorities. In the case of ESRF, intergovernmental agreement with the United Kingdom was difficult to come by (Cramer 2020). These situations are indicative for a more general pattern of recurrent political concerns with the United Kingdom in the context of European politics. This does not only refer to the difficulties that surrounded the initial EEC membership applications of the country in the 1960s, as shown above, but also to a general skeptical attitude of the British government towards the EEC/EU, which climaxed very recently in the planned withdrawal of the United Kingdom as a member of the EU, so-called Brexit (May 2014: 91).

Third, Russia's substantive contribution to several Big Science projects in Europe since the 1990s, such as ESRF, the European XFEL and FAIR, does not only mirror a general reorientation of Russia's national science policy agenda after the end of the Cold War. This dominating role also resonates how the various relations between the EU, its individual member countries and Russia certainly constitute a major cornerstone of the political and diplomatic landscape of post-Cold War Europe that also translates into the realm of large-scale scientific collaboration.

Summarizing these different aspects, it can be argued that the historical dynamics of the investigated Big Science projects mirrored and resonated broader European political controversies and diplomatic struggles. This adds to the general observation that to the extent that politics matter during the establishment, construction and operation of Big Science projects in Europe, these projects can also provide a strong identificatory moment in times of political crisis as well as political impetus to push political collaboration further beyond the realm of science and technology. It can be concluded that these projects possess a structural function as well as a role in forging relationships between countries that come importantly to matter in times of political disarray or crisis and diplomatic uncertainty.

REFERENCES

Ash M G (2010) Wissenschaft und Politik: Eine Beziehungsgeschichte im 20. Jahrhundert. *Archiv für Sozialgeschichte* **50**: 11–46.

Atkinson H (1997) Commentary on the history of ILL and ESRF. In J Krige and L Guzzetti (eds) *History of European scientific and technological cooperation*. European Communities, pp 144–53.

Bacon G (1987) Introduction: The pattern of 50 years. In G Bacon (ed) *Fifty years of neutron diffraction: The advent of neutron scattering*. Adam Hilger, pp 1–10.

Blaauw A (1991) *ESO's early history: The European Southern Observatory from concept to reality*. ESO Publishing.

Blaauw A (1997) History of the European Southern Observatory (ESO). In J Krige and L Guzzetti (eds) *History of European scientific and technological cooperation*. European Communities, pp 109–19.

Cabane L and M Lodge (2018) *Dealing with transboundary crises in the European Union: Options for enhancing effective and legitimate transboundary crisis management capacities*. TransCrisis Working Paper. Available at eprints.lse.ac.uk/91587/1/Lodge_Dealing-with-transboundary.pdf (last accessed October 10, 2019).

Capshew J H and K A Rader (1992) Big science: Price to the present. *Osiris* 2nd series **7**: 3–25.

Cesarsky C and C Madsen (2007) Focussing European astronomy: ESO's in the "comeback" of European astronomy. In A Heck (ed) *Organizations and strategies in astronomy*. Springer, pp 97–113.

Cramer K C (2017) Lightening Europe: Establishing the European Synchrotron Radiation Facility (ESRF). *History and Technology* **33** (4): 396–427.

Cramer K C (2020) *A political history of Big Science: The other Europe*. Palgrave Macmillan.

Crease R (1999) *Making physics: A biography of Brookhaven National Laboratory, 1946–1972*. University of Chicago Press.

Crease R and C Westfall (2016) The New Big Science. *Physics Today* **69** (5): 30–6.

Cross M (2018) *The politics of crisis in Europe*. Cambridge University Press.

Defrance C (2016) France-Allemagne: Une Coopération Scientifique "Privilégiée" en Europe, de l'Immédiat Après-Guerre au Milieu des Années 1980? In C Defrance and

A Kwaschik (eds) *La Guerre Froide et l'Internationalisation des Sciences: Acteurs, Réseaux et Institutions*. CNRS, pp 169–86.
Defrance C and U Pfeil (2013) Die Entwicklung der Deutsch-Französischen Kulturbeziehungen nach dem Ende des Kalten Krieges. In M Koopmann, J Schild and H Stark (eds) *Neue Wege in ein Neues Europa: Die Deutsch-Französischen Beziehungen nach dem Ende des Kalten Krieges*. Nomos, pp 179–98.
Demetriou K N (2015) *The European Union in crisis: Explorations in representation and democratic legitimacy*. Springer.
Dinan D, N Nugent and W E Paterson (2017) *The European Union in crisis*. Palgrave.
European Commission (2013) *Review of the S&T cooperation agreement between the European Union and Russia*. European Commission.
European XFEL (2009) *Convention concerning the construction and operation of a European X-Ray Free-Electron Laser Facility* (signed November 30, 2009).
FAIR (2010) *Convention concerning the construction and operation of a facility for antiproton and ion research in Europe* (signed October 12, 2010).
Habermas J (2012) *The crisis of the European Union: A response*. Polity Books.
Hallonsten O (2012) Continuity and change in the politics of European scientific collaboration. *Journal of Contemporary European Research* **8** (3): 300–19.
Hallonsten O (2014) The politics of European collaboration in Big Science. In M Mayer, M Carpes and R Knoblich (eds) *The global politics of science and technology*, Vol. 2. Springer.
Hallonsten O (2016) *Big Science transformed: Science, politics and organization in Europe and the United States*. Palgrave Macmillan.
Hallonsten O (2020) Research infrastructures in Europe: The hype and the field. *European Review* **28** (4): 617–35.
Hallonsten O and T Heinze (2012) Institutional persistence through gradual adaptation: Analysis of national laboratories in the USA and Germany. *Science and Public Policy* **39** (4): 450–63.
Hallonsten O and T Heinze (2013) From particle physics to photon science: Multidimensional and multi-level renewal at DESY and SLAC. *Science and Public Policy* **40** (5): 591–603.
Hermann A, U Mersists, D Pestre and J Krige (1987) *History of CERN. Volume I: Launching the European Organization for Nuclear Research*. North Holland.
Hirschel E-H, H Prem and G Madelung (2004) *Aeronautical research in Germany: From Lilienthal until today*. Springer.
Hoddeson L, A Kolb and C Westfall (2008) *Fermilab: Physics, the frontier, and megascience*. University of Chicago Press.
Jacrot B (2006) *Des neutrons pour la science: Histoire de l'Institut Laue-Langevin, une coopération internationale particulièrement réussie*. EDP Sciences.
Judt T (2005) *Postwar: A history of Europe since 1945*. Penguin Press.
Kariya N C (2009) European space policy and the construction of a European collective identity. Available at https://lib.dr.iastate.edu/etd/10853 (last accessed October 1, 2019).
Koselleck R and M Richter (2006) Crisis. *Journal of the History of Ideas* **67** (2): 357–400.
Krige J (1992) *The prehistory of ESRO, 1959/60: From the first initiatives to the formation of the COPERS*. ESA Publishing.
Krige J (2003) The politics of European scientific collaboration. In J Krige and D Pestre (eds) *Companion to science in the twentieth century*. Routledge, pp 897–918.

Krige J (2006) *American hegemony and the postwar reconstruction of science in Europe*. MIT Press.

Krige J, A Russo and L Sebesta (2000) *A history of the European Space Agency 1958–1987. Volume I: The story of ESRO and ELDO, 1958–1973*. ESA Publishing.

Krotz U and J Schild (2013) *Shaping Europe: France, Germany, and embedded bilateralism from the Elysée Treaty to twenty-first century politics*. Oxford University Press.

Krzyżanowski M (2019) Brexit and the imaginary of "crisis": A discourse-conceptual analysis of European news media. *Critical Discourse Studies* **16** (4): 465–90.

Kühnhardt L (2008) *European Union – the second founding: The changing rationale of European integration*. Nomos.

May A (2014) *Britain and Europe since 1945*. Taylor and Francis.

Middlemas K (1995) *Orchestrating Europe: The informal politics of the European Union 1973–1995*. Fontana Press.

Moedas C (2016) Science diplomacy in the European Union. *Science and Diplomacy* **5** (1).

Moskovko M, A Astvaldsson and O Hallonsten (2019) Who is ERIC? The politics and jurisprudence of a governance tool for collaborative European research infrastructures. *Journal of Contemporary European Research* **15** (3): 249–68.

Patel K K (2018) *Projekt Europa. Eine kritische Geschichte*. C H Beck.

Pestre D (1997) The prehistory of the Franco-German Laue-Langevin Institute. In J Krige and L Guzzetti (eds) *History of European scientific and technological cooperation*. European Communities, pp 137–43.

Prime Minister of the Russian Federation (2011) *Prime Minister Vladimir Putin holds a session of the Government Commission on High Technology and Innovation in Dubna*. Archive of the Official Site of the 2008–12 Prime Minister of the Russian Federation Vladimir Putin, July 5. Available at http://archive.premier.gov.ru/eng/events/news/15785 (last accessed December 5, 2018).

Rekers J and K Sandell (eds) (2016) *New Big Science in focus*. Lund University Press.

Runciman D (2016) What time frame makes sense for thinking about crises? In P F Kjaer and N Olsen (eds) *Critical theories of crisis in Europe: From Weimar to the euro*. Rowman and Littlefield, pp 3–16.

Rush J (2015) US neutron facility development in the last half-century: A cautionary tale. *Physics in Perspective* **1** (2): 135–55.

Russian Government (2019) Dmitry Medvedev's meeting with Russian scientists working at CERN. Available at http://government.ru/en/news/36968/ (last accessed December 1, 2019).

Saunier G (2001) Prélude à la Relance de l'Europe. Le Couple Franco-Allemand et les Projets de Relance Communautaire Vus de l'Hexagone (1981–1985). In M-T Bitsch (ed) *Le Couple France-Allemagne et les Institutions Européennes: Une Postérité pour le Plan Schuman?* E. Bruylant, pp 463–85.

Schmied H (1990a) The European synchrotron radiation story. *Synchrotron Radiation News* **3** (1): 18–22.

Schmied H (1990b) The European synchrotron radiation story – Phase II. *Synchrotron Radiation News* **3** (6): 22–6.

Seidel R (1992) The origins of the Lawrence Berkeley Laboratory. In P Galison and B Hevly (eds) *Big science: The growth of large-scale research*. Stanford University Press, pp 21–45.

Stahlecker T and H Kroll (2013) *Policies to build research infrastructures in Europe: Following traditions or building new momentum?* Fraunhofer ISI Working Papers

Firms and Region No. R4/2013. Available at www.isi.fraunhofer.de/content/dam/isi/dokumente/ccp/unternehmen-region/2013/ap_r4_2013.pdf (last accessed August 1, 2019).

Timmins G and J Gower (2009) Introduction: Russia and Europe: What kind of partnership? In J Gower, G Timmins and G Robertson of Port Ellen (eds) *Russia and Europe in the twenty-first century: An uneasy partnership*. Anthem Press, pp xxi–xxvi.

Trischler H (2010) Physics and politics: Research and research support in twentieth century Germany in international perspective: An introduction. In H Trischler and M Walker (eds) *Physics and politics: Research and research support in twentieth century Germany in international perspective*. Franz Steiner Verlag, pp 9–18.

van der Bliek J (1996) *ETW, a European resource for the world of aeronautics: The history of ETW in the context of European aeronautical research and development*. ETW Publishing.

van Middelaar L (2019) *Alarums and excursions: Improvising politics on the European stage*. Agenda Publishing.

Warlouzet L (2014) Dépasser la crise de l'histoire de l'intégration Européenne. *Politique Européenne* **44** (2): 98–122.

Weingart P (1999) Scientific expertise and political accountability: Paradoxes of science in politics. *Science and Public Policy* **26** (3): 151–61.

Westfall C (2012) Institutional persistence and the material transformation of the US national labs: The curious story of the advent of the advanced photon source. *Science and Public Policy* **39** (4): 439–49.

Woltjer L (2006) *Europe's quest for the universe*. EDP Sciences.

4. Ever-changing Big Science and Research Infrastructures: Evolving European Union policy

Inga Ulnicane

1. INTRODUCTION

Big Science and Research Infrastructures (RIs) keep changing and evolving along with major organizational, technological and political changes that affect science. A recent comprehensive analysis of the transformation of Big Science of the Cold War era to a new context showed that the "old Big Science" was characterized by large teams and long-term experiments, accelerators for particle collisions and reactors for nuclear research and had a clear military connection (Hallonsten 2016). In contrast, the key features of the "transformed Big Science" include large support organizations, accelerators and reactors for neutron scattering, synchrotron radiation and free electron laser; it focuses on innovation-based (regional economic) growth, sustainability and addressing grand challenges.

This chapter aims to study ongoing changes of Big Science and RIs by looking at evolving European Union (EU) policy. Since the year 2000, important changes have taken place in EU policymaking that affect Big Science and RIs. In particular, this chapter will look at the emergence of EU policy for RIs in a *differentiated integration mode*, EU support for politically motivated large-scale research projects and e-infrastructures, as well as the development of RI policy within the European Research Area (ERA) initiative. To illustrate these EU policy changes, an example of an EU-funded large-scale research initiative, the Human Brain Project (HBP), currently developing, will be used. The main research method in this chapter will be document analysis. Study of public discourse and official announcements – as one of the often used qualitative methods to study Big Science and RIs (see Rüffin, ch 2 in this volume) – will be used to analyze a broad range of documents from EU institutions and expert groups.

Thus, this chapter aims to contribute to this volume by delving into EU-level policy changes affecting Big Science and RIs in Europe. This way, it complements other chapters focusing on policy, such as Bolliger and Griffiths (ch 5 in this volume) analyzing RI roadmaps, and Moskovko (ch 6 in this volume) on the European Research Infrastructure Consortium (ERIC) framework. While there are many factors affecting Big Science and RIs, policy is definitely one of them as it sets out boundary conditions (regulation, coordination mechanisms, funding, etc.) for establishing and sustaining Big Science and RI projects. Due to important changes taking place in EU research policy since 2000, this is an important topic for understanding current contexts in which Big Science and RIs operate in Europe.

This chapter proceeds as follows: first, concepts of Big Science, RIs, international research collaboration and science diplomacy are discussed. Second, different models of European integration – intergovernmental, uniform and differentiated – are discussed to shed light on the evolution of EU policy concerning Big Science and RIs in a differentiated integration mode. Third, changes in EU research policy related to large-scale politically motivated projects, e-infrastructures and the ERA are outlined. Fourth, the example of the HBP is used to illustrate some trends in EU policy changes in this area. Finally, conclusions and questions for future research are presented.

2. CONCEPTUAL FRAMEWORK: HOW TO UNDERSTAND BIG SCIENCE AND RESEARCH INFRASTRUCTURES

This chapter draws on the related concepts of Big Science, RIs, international research collaboration and science diplomacy, and applies them to relevant developments in Europe. While the partly overlapping concepts of Big Science and RIs have been understood and used by researchers and policymakers in many different ways, this chapter follows the common approach outlined for this book by Cramer et al (ch 1 in this volume).

Big Science is understood here according to Hallonsten (2016: 17) as "science made big in three dimensions: big organizations, big machines, and big politics". Big organizations mean the organization of large scientific projects in an industrial manner, or hierarchical structure of large teams formed around a large and costly scientific instrument (Cramer et al, ch 1 in this volume). Big machines refer to the size of scientific instruments, while big politics point to substantial political support needed for large-scale research endeavors.

Many Big Science facilities and projects are also RIs, which in this book is understood as a specific policy concept widely used in the EU (Cramer et al, ch 1 in this volume). The European Commission defines RIs as "facilities, resources and services used by the science community to conduct research and foster innovation". RIs "include: major scientific equipment, resources such as collections, archives or scientific data, e-infrastructures such as data and computing systems, and communication networks" that "can be single-sited (a single resource at a single location), distributed (a network of distributed resources), or virtual (the service is provided electronically)". The Commission categorizes European RIs in three organizational groups: firstly, *intergovernmental* that are established by the member states; secondly, *new pan-European* that are listed in the European Strategy Forum on Research Infrastructures (ESFRI) roadmap, including ERICs; and thirdly, *national RIs of European interest* that receive European support (European Commission 2019a).

After analyzing the 60 RIs identified as important for Europe by the EU institutions, Hallonsten (2020) concludes that the concept of RIs is ill-defined, and that the current policy hype around RIs in Europe "is not matched by any substance on the side of what qualifies as a RI and not, and why". According to him, "Research Infrastructures (RIs) are resources that enable scientific research or development work. They can be open-ended in their use, and they can take a variety of technological shapes and forms including instruments and tools for discovery and experimentation, repositories of data and materials, and vessels for exploration". He highlights the political origins and political usefulness of this term in the context of a recent increase in policy attention that the concept of RIs has received at the EU level.

Many Big Science initiatives and large-scale RIs are international because their scale requires pooling of scientific and material resources from several countries. Such RIs are examples of international research collaborations (Ulnicane 2015a). Scientific research has for centuries been characterized by active (international) collaboration, exchange of knowledge and mobility across borders (Crawford et al 1993). Today, research practices are highly internationalized (Nedeva 2013) and international research collaboration as measured by co-authorships is increasing (Wagner et al 2015) due to factors such as increasing scientific specialization and complexity of research, the need to address cross-border problems and escalating costs of research equipment.

To understand RIs as a specific kind of international research collaboration, it is useful to draw on a distinction made by Wagner (2008) between "top-down" and "bottom-up" international research collaborations. Many scientific collaborations are "bottom up", namely driven and organized by individual researchers. They are typically small-scale initiatives to combine complementary expertise and capabilities to address research questions of mutual interest and solve scientific problems. In contrast, Big Science and RIs typically require large-scale national and international funding. According to

Wagner (2008: 26), "[g]overnment officials typically plan such facilities in discussion with scientists and sink significant investment in their construction before any research ever takes place. The organization of these activities can therefore be considered 'top-down'". Thus, in contrast to small-scale self-organizing international research collaborations among scientists, Big Science and RIs that are international collaborations are more dependent on political and governmental support and steering.

As new Big Science initiatives are launched, and this mode of collaboration and organization expands to further scientific disciplines, debate over its relative merits continues. Among the benefits of Big Science are typically mentioned opportunities to bring together diverse types of expertise across disciplinary, organizational and national boundaries to address complex problems, while bureaucratization and politicization are often highlighted as negative consequences of Big Science (e.g. Vermeulen et al 2010).

Large-scale international scientific collaborations behind Big Science and RI initiatives today are often described by another popular term of *science diplomacy*. A broad understanding of the concept of science diplomacy typically sees it as an intersection between science and technology policy and foreign affairs (Royal Society 2010). A popular definition, originating with the Royal Society (2010: 15), distinguishes between the three dimensions of science diplomacy. First, "science in diplomacy" focuses on informing foreign policy objectives with scientific advice. Second, "diplomacy for science" facilitates international scientific cooperation. Third, "science for diplomacy" uses science cooperation to improve international relations. The second dimension, *diplomacy for science*, is the most relevant for understanding political processes and diplomatic efforts behind establishing and maintaining Big Science and RI projects. Some Big Science and RI initiatives can also be examples of the third dimension, *science for diplomacy*, where science cooperation is seen as a tool to improve diplomatic relations.

The following analysis of changing organization, instrumentation and politics of Big Science and RIs will draw on the above-mentioned concepts.

3. MODES OF EUROPEAN INTEGRATION AND CHANGING BIG SCIENCE AND RESEARCH INFRASTRUCTURES

As the introductory chapter to this book reminds us, the history of Big Science and RIs is closely connected to the history of European integration. European integration processes, institutions and policies have changed over time and "the ever-changing Union" (Egenhofer et al 2011) has also had an impact on Big Science and RIs. This section will look at different modes of European integration – intergovernmental, uniform and differentiated – and their relevance for Big Science and RIs. The intergovernmental mode of integration refers to collaboration among governments outside the EU framework, uniform mode of integration refers to policies and institutions that applies to all EU member states, while differentiated integration includes EU policies in which some EU member states do not participate while some non-members may participate. History and policies of Big Science and RIs in Europe include elements of all these three modes: intergovernmental, uniform and differentiated. Looking at Big Science and RIs from the perspective of different modes of European integration helps to connect developments of Big Science and RIs with long-term processes in European integration.

3.1 From Intergovernmental Cooperation to European Union Policy

Major Big Science initiatives started in the 1950s as intergovernmental cooperation among a number of national governments. At that time uniform EU integration in the field of research was almost non-existent. These intergovernmental initiatives included large-scale research facilities such as the European Organization of Nuclear Research (CERN, established in 1954), the European Southern Observatory (ESO, established in 1962), the European Space Research Organisation (ESRO) and European Launcher Development Organization (ELDO) (1964), the Institut Laue-Langevin (ILL) (1967), the European Molecular Biology Laboratory (EMBL, established in 1974), the European Space Agency (ESA, established in 1975) and the European Synchrotron Radiation Facility (ESRF) (1988) (see Cramer, ch 3 in this volume; D'Ippolito and Rüling, ch 11 in this volume).

These intergovernmental initiatives were "built on ad hoc solutions rather than a coherent political framework and common regulatory standards" (Hallonsten 2014: 31). Diverse organizational formats have been used: while CERN, ESA, ESO and EMBL are international organizations operating according to international rules with permanent international staff, ESRF and ILL are private entities operating as companies according to the legislation of the host country and with national labor law constraints (Papon 2009: 36).

Such lack of coherence and reliance on ad hoc solutions for each new intergovernmental initiative has both advantages and disadvantages. Advantages are flexibility, avoidance of bureaucracy and institutional inertia and allowing each initiative to meet the demands of its specific scientific community at a specific time (Hallonsten 2014: 35; Papon 2004), while disadvantages include lack of transparency, an in-built uncertainty, unpredictability, the need to reinvent legal arrangements and organizational structures for each project and, thus, delays in their realization (Hallonsten 2015). Typically, critical issues involved in setting up such intergovernmental initiatives are site selection, fair return and in-kind contributions (Hallonsten 2012), which often have to be resolved through complex political compromises. Hallonsten (2014: 35) argues that each country joining such initiatives undertakes a multitrack cost-benefit analysis, weighing possible economic, political, diplomatic and reputational gains and losses. According to him, "most countries realize that collaboration is necessary to achieve goals beyond the reach of any one of them, but strong traditions of sovereignty create tension between self-interest and common good, for every partaking country, in every collaboration" (Hallonsten 2014: 35). Krige (2002, 2003) argues that European intergovernmental science organizations aim at both promoting the national interests of the participating countries and contributing to the European integration process, and that they are not undertaken at the expense of self-interest but rather allow countries "the pursuit of one's interests by other means" (Krige 2003: 900).

In parallel with intergovernmental cooperation among member states, EU integration gradually expanded to new policy areas, including research. Major milestones in the development of EU research policy include the establishment of multi-annual EU Framework Programmes (FP) for funding research, in 1984, and the launch of the ERA initiative in 2000. According to the Lisbon Treaty that came into force in 2009, research policy is a shared competence between the EU and the member states, which implies that in this policy area both the EU and its member states are able to legislate and adopt legally binding acts. Major EU policy developments for Big Science and RIs include the establishment of ESFRI in 2002 (Bolliger and Griffiths, ch 5 in this volume) and the approval of the ERIC legal framework in 2009 (Moskovko, ch 6 in this volume; Moskovko et al 2019).

The ERIC framework was established by a Council of the EU regulation in 2009 (European Commission 2019b) to facilitate the joint establishment and operation of research infrastructures of European interest among several member states and associated countries. Additionally, membership of ERICs can include third countries and intergovernmental organizations. The minimum requirement for setting up an ERIC is to have at least one EU member state and two other countries that are either EU member states or associated countries. Member states and associated countries must jointly hold the majority of voting rights in the assembly of members. The headquarters of a consortium should be located either in a member state or in an associated country. The ERIC framework has been used both for establishing new infrastructures and as a new organizational framework for existing research initiatives (Moskovko, ch 6 in this volume).

Thus, the ERIC framework builds on experiences of intergovernmental initiatives and offers EU solutions to some of the difficulties encountered (e.g. lengthy negotiations, reinventing new rules for each facility ad hoc) to ensure a faster and more transparent process of establishing joint research facilities

(European Commission 2014b). However, while the ERIC framework can solve some of the problems it can also create new ones. Some important practical advantages and challenges of this framework were illustrated by the case of the European Social Survey becoming an ERIC in 2013. For the Survey, becoming an ERIC allowed it to tackle the challenge of long-term survival and funding but also led to the decline in the number of participating countries and thus to less comprehensive coverage because a number of countries decided not to join the ERIC or to join only as observers (Duclos Lindstrom and Kropp 2017).

3.2 Differentiated Integration

The ERIC model, that brings together only some EU member states and might include countries from outside the EU, is a case of the differentiated (rather than uniform) mode of European integration. Moreover, it is not the only case of differentiated integration in research policy. The core instrument of the EU research and innovation policy, namely the FP, is also a case of differentiated integration because in addition to the EU member states it also includes a considerable number of associated countries (Fumasoli et al 2015; Langfeldt et al 2012; Lavenex 2009). Sixteen associated countries are participating fully or partially in the most recent FP called Horizon 2020 and are making financial contributions to it (European Commission 2019c). Considering the significance of differentiated integration for RIs, and more generally for EU research policy (Chou and Ulnicane 2015), it is worthwhile to take a closer look at differentiated integration, what is driving it and why initiatives for RIs and research policy develop as cases of differentiated rather than uniform integration.

Differentiated integration characterizes "all those policies, in which the territorial extension of European Union (EU) membership and EU rule validity are incongruent" (Holzinger and Schimmelfennig 2012: 292). Well-known examples of differentiated integration in other policy areas include the Economic and Monetary Union where several EU member states do not participate, and the Schengen zone where some EU members do not participate but some non-members participate (Holzinger and Schimmelfennig 2012). Such initiatives, also known as "multi-speed Europe", "flexible integration", "variable geometry", "Europe à la carte" and "graded membership" (Holzinger and Schimmelfennig 2012; Kölliker 2001; Leruth and Lord 2015; Schimmelfennig 2016), are seen as "an essential and, most likely, enduring characteristic of the EU" (Schimmelfennig et al 2015: 765).

While historically examples of differentiated integration might have been seen as temporary phenomena that eventually will lead to full EU integration, data suggest that differentiation has significantly increased over time

(Schimmelfennig et al 2015: 770). In European studies today, differentiated integration is recognized as "a permanent, organizational principle of the Union, grounded in a need to manage divisions and disagreements that just do not go away" (Leruth and Lord 2015: 758). Thus, differentiation has increasingly become a normal feature of European integration, including in the field of research where it has hardly been studied. This highlights the need to better understand the reasons behind and consequences of differentiated integration.

Schimmelfennig et al (2015: 765) distinguish between vertical and horizontal differentiation. While in their view vertical differentiation means that policy areas have been integrated at different speeds and have reached different levels of centralization over time, horizontal differentiation relates to the territorial dimension and refers to the fact that many integrated policies (including research) are neither uniformly nor exclusively valid in the EU member states. Furthermore, horizontal differentiation can be internal, i.e. some member states do not participate in integration, or it can be external, i.e. some non-members participate in selected EU policies (Schimmelfennig et al 2015: 767).

To explain horizontal differentiation, Schimmelfennig et al (2015: 765) propose two main factors – "interdependence" and "politicization", with interdependence acting as a driver of integration, and politicization acting as an obstacle to uniform integration. Interdependence and politicization vary across policy area, country and time (Leruth 2015; Schimmelfennig et al 2015), and internal horizontal differentiation results from high interdependence and high politicization (Schimmelfennig et al 2015). Interdependence focuses on the benefits of cooperation, while the main indicators of politicization are "mass-level salience and contestation of European integration, the mobilization of Eurosceptic public opinion by Eurosceptic parties and opportunities to voice Eurosceptic opinions in national referendums or elections to the European Parliament" (Schimmelfennig et al 2015: 771). Admitting that their comparative categorical measurement of the conditions of interdependence and politicization is not straightforward and is based on plausibility, Schimmelfennig et al (2015: 775) argue that cases of "high interdependence" are characterized by high transnational exchange, significant economies of scale and important international externalities. They characterize cases of "high politicization", with significant autonomy concerns for governments and identity concerns for nation-state citizens.

Do these two factors – interdependence and politicization – help us to understand drivers behind differentiated integration in the case of RIs, namely why some EU non-members are eager to join ERIC initiatives and a number of member states decide not to join? Can high interdependence and high politicization explain differentiated integration initiatives in the case of RIs? The first explanatory variable – high interdependence – is clearly present. As explained above, RIs on the European level are characterized by an active international research collaboration which is necessary to mobilize the required scientific, material, political and financial resources. However, while the overall interdependence of research activities is high, there can be some variation in the level of interdependence across specific research fields and countries. It has been argued that intergovernmental research funding schemes should be "flexible enough for the requirements of different knowledge areas and societal challenges" (Edler 2012: 331). Thus, it can be expected that differentiated integration in research policy results from the variation of interdependence across different research fields for different countries.

The second explanatory variable – high politicization – is well known in the history of European research integration. Even if the main indicators of politicization in the differentiated integration literature outlined above (e.g. mass-level salience and contestation of European integration, mobilization of public opinion, opportunities to voice Eurosceptic opinions) are not present in the case of RIs, major initiatives in this field – as outlined above – have been surrounded with political debates about national sovereignty versus common good and related questions of *juste retour* and political, economic and diplomatic costs and benefits. Hallonsten (2016: 33–4) refers to the concept of "pork barrel politics" to emphasize the importance of political considerations – e.g. benefits of hosting a major facility (namely political "pork") – in the launch and maintenance of Big Science and large-scale RIs. Thus, according to the differentiated integration approach, it can be expected that a country will join a Big Science or RI initiative such as ERIC in a specific research field if it has high scientific interdependence and low politicization in terms of negative public opinion (e.g. Euroscepticism). Clear political benefits (i.e. "pork"), such as the opportunity to host the project, would facilitate participation.

Thus, the differentiated integration approach can provide a useful lens for studying the dynamics of forming and running initiatives such as ERICs and exploring their benefits and shortcomings in the light of research policy and European integration. However, further operationalization and systematic application of this framework for the cases of RIs goes beyond the scope of this chapter and should be considered for future studies.

4. CHANGES IN EUROPEAN UNION POLICY AFFECTING BIG SCIENCE AND RESEARCH INFRASTRUCTURES

To study developments in EU research policy affecting Big Science and RIs, this chapter adopts the three-dimensional framework of Big Science – Big Organizations, Big Machines and Big Politics – developed by Hallonsten (2016). While this framework was developed to study changes in a different historical context, this chapter adopts it to study ongoing developments along organizational, instrumentation and political dimensions that in this case are EU support for politically motivated large-scale projects and e-infrastructures as well as developing EU policies towards RIs in the context of the ERA. While it is not possible to provide an exhaustive analysis of changes in EU policies, the three above mentioned developments capture a number of important trends affecting Big Science and RIs in Europe.

4.1 Big Organizations: Politically Motivated Large-Scale Research Projects

One novel feature of Big Science in Europe is the emergence of politically motivated large-scale research projects. If traditionally the size of science organizations has been mainly related to epistemic features of scientific disciplines and research fields, then in this new type of organization the size of the project is determined by political decisions. Traditionally, EU science funding has supported a large number of relatively small science projects. In EU policy rhetoric, this has been interpreted as fragmentation of funding, and calls have been made for large-scale EU initiatives to compete with other global science super-powers, especially the United States. The Innovation Union Flagship Initiative of the Europe 2020 strategy launched in 2010 highlighted the importance of partnering as a means to address Europe's insufficient and fragmented research and innovation efforts (European Commission 2014a). Such large-scale initiatives in the 7th FP and Horizon 2020 included industry-led public-private partnerships (joint technology initiatives, contractual public-private partnerships), member state-led public-public partnerships (ERA-NET Cofund, Article 185 initiatives, joint programming initiatives), European Innovation Partnerships, Knowledge and Innovation Communities and Future Emerging Technologies (FET) flagships.

This section takes a closer look at the FET Flagships, which in comparison to other large-scale partnerships are envisaged to be science-driven research initiatives that are led by scientific communities in shaping their programs and building up industrial partnership over time (European Commission 2014a). The plan to identify and launch "two or three bold new FET research flagship initiatives which will drive large multidisciplinary research community efforts towards foundational breakthroughs at the frontier of ICT [information and communication technology]" was announced in the communication "Moving the ICT frontiers" (European Commission 2009). This document was published during the times of economic crisis, when EU rhetoric presented increased investments in research and innovation as sources to boost Europe's competitiveness and to facilitate renewed growth (Ulnicane 2016b).

Table 4.1 Evolution of FET Flagships

Year	FET Flagship development
2009	Communication "Moving the ICT frontiers" suggests a new funding instrument: FET Flagships
2010	Open consultation with the scientific community on initial ideas Call for FET Flagships published and 21 proposals received
2011	Selection of six pilot projects after evaluation of initial proposals
2012	Six pilot projects submit their flagship proposals
2013	Two winners announced: Graphene and Human Brain Project Start of the projects
2014	Staff working document on FET Flagships
2015	Signature of Framework Partnership Agreements under Horizon 2020 (8th FP)
2017	Publication of FET Flagships interim evaluation
2018	Launch of the third FET Flagship Quantum
late 2018/ early 2019	During Horizon Europe negotiations (9th FP), plans for future FET Flagships abandoned Unclear future for six new pilot projects that were selected in 2018

Source: Own compilation based on European Commission (2013, 2019d); Kupferschmidt (2019).

Building on the well-regarded FET program for high risk research, the European Commission called for a doubling of its investment in transformative foundational research and FET by 2015, and the launch of FET Flagships was part of this agenda (European Commission 2009). While the main focus of FET Flagships is on ICT research, they are interdisciplinary. The FET Flagships "run typically for a period of 10 years and mobilise hundreds of researchers across Europe with an overall support of around EUR 1 billion" (European Commission 2019d). While these initiatives are often known as "one billion projects", actual funding from the EU FP can be less than half of that expecting that the rest has to be attracted from industry, national government and other sources.

After several years of consultations and competitive multistage selection processes, in 2013, the first two FET Flagships – Graphene and the HBP – were launched (for the timeline of the development of FET Flagships see Table 4.1). The third flagship Quantum Technology was launched in 2018.

According to the European Commission (2014a),

> flagships are visionary, science-driven, large-scale initiatives addressing grand scientific and technological (S&T) challenges. They are long-term initiatives bringing together excellent research teams across various disciplines having a unifying goal and an ambitious research roadmap on how to achieve it. Flagships aim at transformational impacts on science and technology, delivering a key competitive advantage for European industry and substantial benefits for society.

Flagships are expected to bring future growth and competitiveness and "establish Europe as a global leader in their domain" (European Commission 2014a). Under the Horizon 2020 Programme, which is organized according to the three priorities of Excellent Science, Industrial Leadership and Societal Challenges (Ulnicane 2015b, 2016a), FET Flagships are funded under the Excellent Science Pillar (European Commission 2019e). However, as the framing of FET Flagship aims in the EU documents suggests they are expected to deliver on multiple objectives of excellence, relevance and innovation.

During the preparations for the next EU FP called Horizon Europe (2021–27), it was decided to abandon the plans for future FET Flagships (Kupferschmidt 2019) despite the fact that six new flagship pilots have been selected and awarded €1 million each for preparing their proposals (Abbott 2019). The ending of the FET Flagship instrument has been explained by the need to streamline too many different funding instruments and funding approaches (Kupferschmidt 2019).

While the three FET Flagships continue to operate, it is important to analyze what can be learned from this large-scale politically motivated science funding experiment for future organization and funding of research. In cases like this, when the size and length of the project is motivated by the political agenda on competitiveness, the two dimensions of Big Organizations and Big Politics become closely intertwined.

4.2 Big Machines: Support for e-Infrastructures

The development of instrumentation of Big Science and RIs are undergoing important changes. In addition to traditional Big Machines such as accelerators and reactors, new types of scientific instrumentation and apparatus are emerging including digital research infrastructures (Olson et al 2008; see also Franssen, ch 7 in this volume). The EU policy is actively promoting these new types of digital infrastructures that it calls "e-Infrastructures" and that "provide digital-based services and tools for data- and computing-intensive research in virtual and collaborative environments" and that the Commission sees as being "key in future development of research infrastructures" (European Commission 2019f). Against the background of discourse of data revolution, the European Commission has launched and funded a number of e-infrastructure initiatives to support European super-computers and the European Open Science cloud.

The need to support these new types of instruments was highlighted already in the ERA initiative that recognized the need for electronic infrastructures such as data repositories and high-speed networks (European Commission 2007) and ICT-based e-infrastructures "that enable increasingly prevalent data-intensive collaborative research by geographically dispersed teams"

(European Commission 2012). Vision for the European scientific e-infrastructures for 2030 was set out by the High Level Expert Group on Scientific Data (European Commission 2010a) framing it as follows: "Our vision is a scientific e-infrastructure that supports seamless access, use, re-use, and trust of data. In a sense, the physical and technical infrastructure becomes invisible and the data themselves become the infrastructure – a valuable asset, on which science, technology, the economy and society can advance". These new types of instrumentation come with new organizational and political aspects that are not present in the cases of traditional Big Machines. One such question is about the ethical, legal and social aspects of Big Science and RIs which are particularly relevant in the case of data infrastructures of life sciences that rely on patient data (Goisauf et al 2019).

4.3 Big Politics: ERA and RIs

If historically RIs in Europe were established as intergovernmental initiatives outside the EU framework, then the main developments towards EU policy on RIs started within the ERA initiative launched in 2000. EU policy towards RIs evolved gradually in parallel to the development of the ERA initiative (see Table 4.2).

As Table 4.2 demonstrates, since the launch of the ERA, the EU policy towards RIs has evolved from a general recognition of the need to define a common approach to very specific actions such as the launch of the legal framework and quantitative targets for the construction of RIs.

The launch of the ERA initiative in 2000 was motivated by concerns about the EU's competitive position vis-à-vis the United States and Japan as well as about the lack of coordination between national and EU research policies (Ulnicane 2015b). To address this, the ERA initiative outlined action lines including more coherence and common approaches to investments and human resources. Among actions, the initial ERA document called for a definition of a European approach to RIs (European Commission 2000). The document stated that already for several years the EU had been operating a program of support for RIs which had provided support for transnational access to facilities, for the development of new instruments and equipment, and for cooperation projects designed to improve the interoperability of installations and the complementarity of their activities. It called for going a step further and developing "a European approach to infrastructures, covering both the creation of new installations, the functioning of existing ones and access to them" (European Commission 2000).

The idea that RIs cannot be dealt with effectively at national level was repeated in an early review of the ERA (European Commission 2002), which reinforced the idea that "research infrastructures top the list of areas where

Table 4.2 Evolution of EU policy towards Research Infrastructures within the ERA initiative

Year	Developments in ERA initiative	Relevance for RIs
2000	Communication "Towards European Research Area"	Need to define a European approach to research infrastructures
2002	Communication "The European Research Area: Providing a new momentum"	Creation of ESFRI
2007	Green Paper "The European Research Area: New perspectives"	Developing world-class research infrastructures, including an appropriate legal structure
2008	Reports from the expert groups set up following the Green Paper	Report of the ERA Expert Group "Developing World-Class Research Infrastructures for the European Research Area"
2009	Lisbon Treaty comes into force with objective of achieving ERA and shared competence for research	Council decision on the ERIC framework
2010		Report of the (second) Expert Group on Research Infrastructures: "A vision for strengthening world-class research infrastructures in the ERA"
2010	Europe 2020 and Innovation Union; deadline to complete ERA by 2014	Innovation Union target: by 2015, launch or complete the construction of 60 percent of the prioritized RIs identified by ESFRI
2012	ERA reform agenda	Effective investment in and use of RIs included in the ERA priorities
2014	Launch of ERA monitoring via regular Progress Reports	Monitoring of ESFRI roadmaps and projects
2015	ERA roadmap 2015–20	Optimal use of public investments in RIs

Source: Own compilation based on Ulnicane (2015b).

a European approach is called for, given the levels of funding and the need for them to be given the means to ensure they are able to provide services on a European scale". The main achievement in the area of RIs reported in this document was the creation of ESFRI in 2002 "to facilitate the emergence of a European policy on the development and use of research infrastructures in Europe, as well as multilateral initiatives in this field" (European Commission 2002).

The ERA Green Paper in 2007, that launched a broader public, stakeholder and expert involvement with the ERA, introduced the discourse of "world-class research infrastructures" (European Commission 2007). One of the six features that the ERA should have, according to the Green Paper, was "world-class research infrastructures, integrated, networked and accessible to research teams from across Europe and the world, notably thanks to new generations of electronic communication infrastructures" (European Commission 2007). Steps towards developing world-class infrastructures included building on the ESFRI roadmap (first published in 2006), making the most of all sources of funding, an appropriate legal structure and further developing electronic infrastructures in Europe and the world.

Following up on the Green Paper, the European Commission established seven Expert Groups, one of which it dedicated to RIs. In 2008 it produced the report "Developing world-class Research Infrastructures for the European Research Area (ERA)" which highlighted the role of RIs for Europe's competitiveness in basic and applied research as well as in generating ideas for industrial, societal and political applications. Among concrete steps, the report mentioned the creation of a legal framework for pan-European RIs. This recommendation was implemented in 2009, when the Council of the EU adopted the ERIC legal framework. In 2010, the second expert group released its report with further recommendations for strengthening RIs in the ERA (European Commission 2010b).

Moreover, in 2010 the Lisbon Strategy was replaced by the Europe 2020 strategy for smart, sustainable and inclusive growth. One of the so-called Europe 2020 Flagship Initiatives was the Innovation Union (European Commission 2010c), which set a deadline for completing the ERA in 2014 as well as a number of specific targets. One of these targets prescribed that "by 2015, Member States together with the Commission should have completed or launched the construction of 60% of the priority European research infrastructures currently identified by the European Strategy Forum on Research Infrastructures (ESFRI)" (European Commission 2010c).

In light of the deadline for completing the ERA in 2014 and other targets set in the Innovation Union Flagship Initiative, in 2012 the European Commission launched the ERA reform agenda (European Commission 2012). A focus on effective investment in and use of research infrastructures was included in one of the ERA priorities of "Optimal transnational co-operation and competition". To facilitate implementation of the ERA priorities, in 2014 an ERA monitoring process was launched with regular ERA Progress Reports which also monitored the progress on RIs. According to the 2018 ERA Progress Report: "As of 2018, over half of the ERA countries had roadmaps in place together with ESFRI research infrastructures, but only a third of them had also identified funding needs. However, more cooperation and synchronization of national procedures (for setting priorities, monitoring and long-term funding) is needed to make the European research infrastructure ecosystem more robust and increase the effectiveness of public investments in this area" (European Commission 2019g). The top priorities of the ERA Roadmap 2015–20 include "making optimal use of public investments in Research Infrastructures (RIs) by setting national priorities compatible with the European Strategy Forum on Research Infrastructures (ESFRI) priorities and criteria taking full account of long term sustainability" (Council of the European Union 2015). Among the points for possible future consideration, the roadmap highlights "the need to step up the efforts in the area of the Research Infrastructures, including the e-infrastructures, and UNDERLINES that the use of the European Structural and Investment Funds for these purposes should be encouraged, where appropriate" (Council of the European Union 2015).

To sum up, the EU policy towards RIs was largely developed within the ERA initiative, demonstrating that the EU includes RIs among the core priorities of European integration in science. Over time this policy developed from broad statements about the need for a common European approach, to concrete actions and decisions. Today EU policy for RIs is largely developed within the ESFRI. Currently, the role of the ERA initiative for RIs is largely limited to monitoring, with results published in the ERA Progress Reports. Overall since the controversial deadline to complete the ERA by 2014 (Ulnicane 2015b, 2016c), the role of the ERA initiative in EU research policy has diminished. However, in 2020 it is planned to launch a new strategy for ERA. The EU policy discourse about RIs has shifted from previous grand ambitions of "world-class infrastructures" (European Commission 2007, 2008, 2010a, 2010b) to a more pragmatic recent focus on "sustainable research infrastructures" (European Commission 2017a).

5. EXAMPLE OF CHANGING EUROPEAN UNION POLICY: THE HUMAN BRAIN PROJECT

To illustrate changes in EU policy towards Big Science and RIs (discussed in Section 4), this section looks at the HBP. It is one of the two initial FET Flagships (see Section 4.1) launched in 2013 to develop an RI for scientific and industrial researchers to study neuroscience, computing and brain-related medicine. After expiration of its current FET Flagship funding in 2023, it is planning to establish itself as an RI initiative called EBRAINS. Preparations for developing a sustainable RI involve a number of diverse activities from preparing services and building user communities to establishing legal entity and ensuring sustainable funding sources.

The HBP is one of the largest science projects ever funded by the EU, bringing together some 500 researchers and engineers from different scientific disciplines including neuroscience, computing and social sciences. They are based in approximately 120 universities, research institutes and hospitals in some 20 countries. While popularly known as a "one billion project", actual funding expected from the FPs over ten years is around €400 million, and the rest of the funding is expected to be acquired from other sources.

5.1 Big Organization: From FET Flagship to Sustainable Research Infrastructure

The EU policy discourse explaining the importance of the HBP states that "understanding the human brain is one of the greatest challenges facing 21st century science. By rising to the challenge, profound insights into what it means to be human can be gained. New treatments for brain diseases may also become possible, and new knowledge may enable revolutionary new computing technologies" (European Commission 2017b).

Ideas for the HBP originate from the Swiss Blue Brain project that aims to reverse-engineer the mammalian brain. At the beginning, the aim of the HBP, championed by its initial leader Henry Markram, was to build a realistic computer model of the brain by 2023. Already in the HBP proposal in 2012, it was mentioned that building and simulating such models will require infrastructure and ICT platforms branded as infrastructure (Human Brain Project 2012). The European Commission's press release presented the HBP as a "facility", stating that it "will create the world's largest experimental facility for developing the most detailed model of the brain, for studying how the human brain works and ultimately to develop personalised treatment of neurological and related diseases" (European Commission 2013).

Interestingly, initial leaders of this highly risky ICT and neuroscience project aiming to simulate the human brain compared the HBP with well-known successful traditional large-scale initiatives and facilities including CERN, Higgs boson, the Human Genome Project, National Aeronautics and Space Administration, space shuttles and telescopes (see e.g. Frackowiak 2014; Honigsbaum 2013). These comparisons can be seen as attempts to make the HBP more acceptable and understandable as well as to signal the level of ambition.

When the highly risky grand vision came under criticism, the aim to turn the HBP into an RI moved to the forefront as a safer objective for the project's continuation. Less than a year after the HBP was launched, an open letter attacking HBP's science and organization gathered more than 800 signatures of neuroscientists (Theil 2015; see also Mahfoud 2020). Additionally, signatories were concerned that the financing of HBP would drain all the EU and national funds on neuroscience research for the duration of the project (European Commission 2017b). Similar concerns that funding of large digital infrastructures would limit resources for other epistemic approaches can be seen also in other scientific disciplines including humanities (see Franssen, ch 7 in this volume). Major scientific and structural reorganization of the

Table 4.3 Timeline: from the Human Brain Project to EBRAINS research infrastructure

Year	Developments
2013	The Human Brain Project was launched following a competitive selection process
2014	Open letter and mediation process; 1st technical project review
2015	Framework Partnership Agreement includes White Paper on Infrastructure
2016	HBP ICT platforms made available to the public
2018	Start of Computational e-Infrastructure FENIX for EU Neuroscience
2019	Establishment of the RI legal entity
2023	FET Flagship funding ends/continuation as EBRAINS RI expected

Source: Own compilation partly based on Human Brain Project (2019a).

HBP followed (with a close involvement of the European Commission) and the main aim of computing the brain was changed to building an RI (for the timeline of the HBP, see Table 4.3). Although the main aim of the project was revised in 2014 and 2015, several years later in the popular press and mass media the HBP is still associated with a vision of building a computer model of the human brain (e.g. Marsh 2019; O'Connell 2017).

The mediation process and the first technical project review that followed the open letter led to important changes including one of the project leadership. The new objectives of the HBP were outlined in the Framework Partnership Agreement (Human Brain Project 2015). The six closely related HBP objectives are (1) to create and operate a European scientific research infrastructure for brain research, cognitive neuroscience and other brain-inspired sciences; (2) to gather, organize and disseminate data describing the brain and its disease; (3) to simulate the brain; (4) to build multiscale scaffold theory and models for the brain; (5) to develop brain-inspired computing, data analytics and robotics; and (6) to ensure that the HBP's work is undertaken responsibly and that it benefits society (Human Brain Project 2015).

The Framework Partnership Agreement also included the White Paper "Transforming the Human Brain Project Platforms into a Community-Driven Infrastructure for Brain Research", which also set out a roadmap for RI development. This focused more specifically on what we here could call instrumentation, and what will be discussed below.

5.2 Big Machines: ICT Research Platforms

The HBP is building the EBRAINS RI for collecting, analyzing, sharing, integrating and modelling data about the brain. At the core of this RI are six ICT platforms launched in 2016: Neuroinformatics (access to shared brain data),

Brain Simulation (replication of brain architecture and activity on computers), High Performance Analytics and Computing (providing the required computing and analytics capabilities), Medical Informatics (access to patient data, identification of disease signatures), Neuromorphic Computing (development of brain-inspiring computing) and Neurorobotics (use of robots to test brain simulation) (Human Brain Project 2019b). Thus, platforms provide access to hardware, software, simulation environments and data sets.

The promise is that "the Platforms will enable new kinds of collaborative research to be performed in neuroscience, medicine and computing. The prototype tools, hardware systems and initial data sets are designed to enable faster and more efficient research techniques in, for example, modelling, in silico experimentation, or data analysis" (European Commission 2017b). Additionally, from 2018 to 2023, the five leading European super-computing centers that form the HBP High Performance Analytics and Computing platform have launched the Interactive Computing E-Infrastructure ICEI/FENIX that provides computing and data services for the HBP as well as for external users.

Key challenges for building research infrastructure include attracting diverse user communities (beyond HBP) and ensuring sustainability and coherence of the infrastructure. To engage the user community and develop collaborations, in 2018 the HBP announced a voucher scheme that gives access to HBP platforms and engineers to solve specific scientific problems. According to the vision for the future of the HBP, "the brain research infrastructure will unify the individual components into one cloud-based super-structure, facilitating the exchange of knowledge, data, models, and algorithms within HBP as well as between the HBP and the 'outside world'" (Amunts et al 2019).

Thus, a major effort to develop research infrastructure in this case is undertaken within a fixed-term project without guaranteed future funding. This leads to the next topic: the Big Politics of the HBP.

5.3 Uncertainty of Big Politics

So far Big Politics has worked for and against the HBP. Initially, its scientific vision of modeling the human brain received major political support at the EU level in terms of being awarded FET Flagship. However, very soon politics backfired when this high-profile project triggered major critique from the scientific community and in response to the open letter the HBP was reorganized.

The international science context has been an important element in the EU's support for the HBP. Already initial EU policy documents on the first two FET Flagships mention that "understanding the human brain is a global challenge" (European Commission 2014a) and refer to other major brain initiatives in the world including the United States, Canada, Japan, China, Korea and Australia.

In popular press, international focus on neuroscience has also been framed as a "brain race" (see e.g. Honigsbaum 2013). In 2017, the HBP together with other major initiatives launched an International Brain Initiative to advance ethical neuroscience research through international collaboration and knowledge sharing (see also Savage 2019). Here we can see also elements of science diplomacy: EU scientists interact and exchange ideas with scientists in the United States and China despite the fact that this often sensitive research takes place under very different regulatory and ethical regimes that among other issues affect treatment of personal data and animal research.

Building and maintaining a sustainable research infrastructure after the FET Flagship (i.e. after 2023), e.g. by being included in the ESFRI roadmap, requires mobilization of major political support also at the national level. This means that not all initial project partners might continue their participation in the RI as has been experienced by other initiatives such as the European Social Survey discussed above (Duclos Lindstrom and Kropp 2017). The differentiated integration approach (Section 3.2) suggests that countries will join differentiated initiatives when interdependence is high and politicization is low. International interdependence is relatively new in neuroscience and ICT, which traditionally have not been examples of Big Science. However, a more recent trend towards major brain initiatives around the world and international collaboration between them might increase motivation for national governments to collaborate internationally in this field.

Low politicization in this case is a more uncertain issue due to the HBP's history. Moreover, the need to acquire national support for participation in an infrastructure initiative implies that composition of countries in future infrastructure initiatives might differ from the current national composition of the HBP, thus affecting the coherence of the infrastructure that is being built.

6. SUMMARY AND OPEN QUESTIONS

Since the year 2000, the EU has launched major policy agendas focusing on increasing its global competitiveness and improving coordination and collaboration among its member states. In response to the financial crisis in 2008–2009, the EU reinforced these policies to stimulate economic growth. Big Science and RIs are part of this broad EU policy agenda.

This chapter described some ongoing changes in EU policy affecting Big Science and RIs and revealing the evolving role of EU policy on the closely interconnected new developments in three dimensions – Big Organizations, Big Machines and Big Politics – in the EU. Political interest to fund large-scale projects increases influence of political factors on Big Organizations. The EU support for e-infrastructures, data repositories and super-computers, facilitated by the rhetoric of data revolution, enables the development of new types of instruments and apparatus in parallel to more traditional Big Machines. The Big Politics of the ERA put RIs at the core of European integration in research. The choice to support RIs in a differentiated integration mode, that brings together selected member states and non-members, provides flexibility but also presents challenges for supranational initiatives.

These developments and the example of the HBP draws attention to a number of open questions about ongoing and future changes and transformations. Are political and scientific factors complementary or contradictory in the case of the launch and governance of large-scale research projects? What lessons can be learned about the governance and effectiveness of such projects? What is the evidence from large-scale research projects on their contribution to fulfilling their diverse objectives of achieving excellence, tackling societal challenges, accomplishing industrial collaboration and contributing to international competitiveness? How can new e-infrastructures be made sustainable? What roles do user communities and other stakeholders play in the development and maintenance of e-infrastructures? Are large-scale digital infrastructures promoted at the expense of other epistemic approaches? Is there a trend towards turning fixed-term scientific projects to become sustainable research infrastructures and is today's science system ready for it? Should EU policymakers consider going beyond differentiated integration-type arrangements in this area, and develop a supranational framework for setting up RIs?

NOTE

ACKNOWLEDGMENTS

This chapter has benefited from feedback on presentations of earlier drafts at the book workshop in Lund, Sweden, 2019 and the General Conferences of the European Consortium for Political Research in 2018 in Hamburg, Germany and in 2019 in Wroclaw, Poland. Useful recommendations from Tara Mahfoud, Christine Aicardi, Bernd Carsten Stahl and the editors of this book are gratefully acknowledged. The research reported in this chapter has received funding from the EU's Horizon 2020 Research and Innovation Programme under Grant Agreements No. 720270 (HBP SGA1), No. 785907

(HBP SGA2) and No. 945539 (HBP SGA3). The author has been a researcher within the HBP's Ethics and Society team since December 2017.

REFERENCES

Abbott A (2019) Europe's next €1-billion science projects: Six teams make it to final round. *Nature* **566**: 164–5.
Amunts K, A C Knoll, T Lippert, C M A Pennartz, P Ryvlin, A Destexhe, V K Jirsa, E D'Angelo and J G Bjaalie (2019) The Human Brain Project: Synergy between neuroscience, computing, informatics, and brain-inspired technologies. PLoS Biology **17** (7): e3000344.
Chou M-H and I Ulnicane (2015) New horizons in the Europe of knowledge. *Journal of Contemporary European Research* **11** (1): 4–15.
Council of the European Union (2015) Draft conclusions on the European Research Area Roadmap 2015–2020, May 19. Available at http://data.consilium.europa.eu/doc/document/ST-8975-2015-INIT/en/pdf (last accessed August 25, 2019).
Crawford E, T Shinn and S Sorlin (eds) (1993) *Denationalizing science: The contexts of international scientific practice*. Springer.
Duclos Lindstrom M and K Kropp (2017) Understanding the infrastructure of European Research Infrastructures: The case of the European Social Survey (ESS-ERIC). *Science and Public Policy* **44** (6): 855–64.
Edler J (2012) Toward variable funding for international science. *Science* **338** (6105): 331–2.
Egenhofer C, P M Kaczynski, S Kurpas and L Schaik (2011) *The ever-changing Union: An introduction to the history, institutions and decision-making processes of the European Union*, 2nd ed. Centre for European Policy Studies.
European Commission (2000) *Towards a European Research Area*. Communication COM(2000) 6.
European Commission (2002) *The European Research Area: Providing new momentum*. Communication COM(2002) 565.
European Commission (2007) *The European Research Area: New perspectives*. Green Paper. COM(2007) 161.
European Commission (2008) *Developing world-class research infrastructures for the European Research Area (ERA)*. Report of the ERA Expert Group.
European Commission (2009) *Moving the ICT frontiers: A strategy for research on future and emerging technologies in Europe*. Communication COM(2009) 184.
European Commission (2010a) *Riding the wave: How Europe can gain from the rising tide of scientific data*. Final Report of the High Level Expert Group on Scientific Data.
European Commission (2010b) *A vision for strengthening world-class research infrastructures in the ERA*. Report from the Expert Group on Research Infrastructures.
European Commission (2010c) *Europe 2020 Flagship Initiative Innovation Union*. Communication COM(2010) 546.
European Commission (2012) *A reinforced European Research Area partnership for excellence and growth*. Communication COM(2012) 392.
European Commission (2013) Graphene and the Human Brain Project win largest research excellence award in history. Available at https://ec.europa.eu/digital-single-market/en/news/graphene-and-human-brain-project-win-largest-research-excellence-award-history (last accessed December 6, 2019).

European Commission (2014a) *FET Flagships: A novel partnering approach to address grand scientific challenges and to boost innovation in Europe*. Commission Staff Working Document SWD(2014) 283 final.
European Commission (2014b) *Report from the Commission to the European Parliament and the Council on the Application of Council Regulation (EC) No 723/2009 of 25 June 2009 on the Community legal framework for a European Research Infrastruicture Consortium (ERIC)*. Communication COM(2014) 460.
European Commission (2017a) *Sustainable European research infrastructures*. Commission Staff Working Document SWD(2017) 323.
European Commission (2017b) *FET Flagships interim evaluation*. Final Report.
European Commission (2019a) *European Research Infrastructures*. Available at https://ec.europa.eu/info/research-and-innovation/strategy/european-research-infrastructures_en (last accessed February 13, 2019).
European Commission (2019b) *European Research Infrastructure Consortium (ERIC)*. Available at https://ec.europa.eu/info/research-and-innovation/strategy/european-research-infrastructures/eric_en (last accessed December 6, 2019).
European Commission (2019c) *Associated countries*. Available at https://ec.europa.eu/research/participants/data/ref/h2020/grants_manual/hi/3cpart/h2020-hi-list-ac_en.pdf (last accessed December 6, 2019).
European Commission (2019d) *FET Flagships*. Available at https://ec.europa.eu/digital-single-market/en/fet-flagships (last accessed February 15, 2019).
European Commission (2019e) *Horizon 2020 FET Flagships*. Available at https://ec.europa.eu/programmes/horizon2020/en/h2020-section/fet-flagships (last accessed August 26, 2019).
European Commission (2019f) *e-Infrastructures*. Available at https://ec.europa.eu/digital-single-market/en/e-infrastructures (last accessed November 24, 2019).
European Commission (2019g) *ERA Progress Report 2018*. Communication COM(2019) 83.
Frackowiak R (2014) Defending the grand vision of the Human Brain Project. *New Scientist* July 16.
Fumasoli T, Å Gornitzka and B Leruth (2015) A multi-level approach to differentiated integration: Distributive policy, national heterogeneity and actors in the European Research Area. *ARENA Working Paper 2*. Available at www.sv.uio.no/arena/english/research/publications/arena-working-papers/2015/wp2-15.html (last accessed January 10, 2020).
Goisauf M, G Martin, H B Bentzen, I Budin-Ljøsne, L Ursin, A Durnova et al (2019) Data in question: A survey of European biobank professionals on ethical, legal and societal challenges of biobank research. *PLoS ONE* **14** (9): e0221496.
Hallonsten O (2012) Continuity and change in the politics of European scientific collaboration. *Journal of Contemporary European Research* **8** (3): 300–19.
Hallonsten O (2014) The politics of European collaboration in Big Science. In M Mayer, M Carpes and R Knoblich (eds) *The global politics of science and technology. Vol. 2: Perspectives, cases and methods*. Springer, pp 31–46.
Hallonsten O (2015) Unpreparedness and risk in Big Science policy: Sweden and the European Spallation Source. *Science and Public Policy* **42** (3): 415–26.
Hallonsten O (2016) *Big science transformed: Science, politics and organization in Europe and the United States*. Palgrave Macmillan.
Hallonsten O (2020) Research Infrastructures in Europe: The hype and the field. *European Review* **28** (4): 617–35.

Holzinger K and F Schimmelfennig (2012) Differentiated integration in the European Union: Many concepts, sparse theory, few data. *Journal of European Public Policy* **19** (2): 292–305.
Honigsbaum M (2013) Human Brain project: Henry Markram plans to spend €1bn building a perfect model of the human brain. *Guardian* October 15.
Human Brain Project (2012) *The Human Brain Project*. A report to the European Commission.
Human Brain Project (2015) *Framework Partnership Agreement*.
Human Brain Project (2019a) Human Brain Project grant structure. Available at www.humanbrainproject.eu/en/about/human-brain-project-ec-grants/ (last accessed December 6, 2019).
Human Brain Project (2019b) Short overview of the Human Brain Project. Available at www.humanbrainproject.eu/en/about/overview/ (last accessed December 6, 2019).
Kölliker A (2001) Bringing together or driving apart the union? Towards a theory of differentiated integration. *West European Politics* **24** (4): 125–51.
Krige J (2002) The birth of EMBO and the difficult road to EMBL. *Studies in History and Philosophy of Science Part C: Studies in History and Philosophy of Biological and Biomedical Sciences* **33** (3): 547–64.
Krige J (2003) The politics of European scientific collaboration. In J Krige and D Pestre (eds) *Companion to science in the twentieth century*. Routledge, pp 897–918.
Kupferschmidt K (2019) Europe abandons plans for "flagship" billion-euro research projects. *Science* May 14.
Langfeldt L, H Godø, Å Gornitzka and A Kaloudis (2012) Integration modes in EU research: Centrifugality versus coordination of national research policies. *Science and Public Policy* **39** (1): 88–98.
Lavenex S (2009) Switzerland in the European Research Area: Integration without legislation. *Swiss Political Science Review* **15** (4): 629–51.
Leruth B (2015) Operationalizing national preferences on Europe and differentiated integration. *Journal of European Public Policy* **22** (6): 816–35.
Leruth B and C Lord (2015) Differentiated integration in the European Union: A concept, a process, a system or a theory? *Journal of European Public Policy* **22** (6): 754–63.
Mahfoud T (2020) Visions of unification and integration: Building brains and communities in the European Human Brain Project. *New Media and Society*.
Marsh H (2019) Can man ever build a mind? *Financial Times* January 10.
Moskovko M, A Astvaldsson and O Hallonsten (2019) Who is ERIC? The politics and jurisprudence of a governance tool for collaborative Research Infrastructures. *Journal of Contemporary European Research* **15** (3): 249–68.
Nedeva M (2013) Between the global and the national: Organising European science. *Research Policy* **42** (1): 220–30.
O'Connell M (2017) *To be a machine: Adventures among cyborgs, utopians, hackers, and the futurists solving the modest problem of death*. Granta.
Olson G M, A Zimmerman and N Bos (eds) (2008) *Scientific collaboration on the internet*. MIT Press.
Papon P (2004) European scientific cooperation and Research Infrastructures: Past tendencies and future prospects. *Minerva* **42** (1): 61–76.
Papon P (2009) Intergovernmental cooperation in the making of European research. In H Delanghe, U Muldur and L Soete (eds) *European science and technology policy: Towards integration or fragmentation?* Edward Elgar Publishing, pp 24–43.

Royal Society (2010) *New frontiers in science diplomacy: Navigating the changing balance of power*. Royal Society.

Savage N (2019) Brain work. Large-scale national research projects hope to reveal the secrets of the human brain. *Nature* **574**: 49–51.

Schimmelfennig F (2016) Good governance and differentiated integration: Graded membership in the European Union. *European Journal of Political Research* **55** (4): 789–810.

Schimmelfennig F, D Leuffen and B Rittberger (2015) The European Union as a system of differentiated integration: Interdependence, politicization and differentiation. *Journal of European Public Policy* **22** (6): 764–82.

Theil S (2015) Why the Human Brain Project went wrong – and how to fix it. *Scientific American* October 1.

Ulnicane I (2015a) Why do international research collaborations last? Virtuous circle of feedback loops, continuity and renewal. *Science and Public Policy* **42** (4): 433–47.

Ulnicane I (2015b) Broadening aims and building support in science, technology and innovation policy: The case of the European research area. *Journal of Contemporary European Research* **11** (1): 31–49.

Ulnicane I (2016a) "Grand challenges" concept: A return of the "big ideas" in science, technology and innovation policy? *International Journal of Foresight and Innovation Policy* **11** (1–3): 5–21.

Ulnicane I (2016b) Research and innovation as sources of renewed growth? EU policy responses to the crisis. *Journal of European Integration* **38** (3): 327–41.

Ulnicane I (2016c) Research paper on the European Research Area initiative and the free circulation of knowledge. In C Salm and T Zandstra (eds) *European Research Area: Cost of non-Europe report*. European Parliamentary Research Service, pp 19–50.

Vermeulen N, J N Parker and B Penders (2010) Big, small or mezzo? *EMBO Reports* **11** (6): 420–3.

Wagner C S (2008) *The new invisibile college: Science for development*. Brookings Institutions Press.

Wagner C S, H Park and L Leydesdorff (2015) The continuing growth of global cooperation networks in research: A conundrum for national governments. *PLoS ONE* **10** (7): e0131816.

5. The introduction of ESFRI and the rise of national Research Infrastructure roadmaps in Europe

Isabel K. Bolliger and Alexandra Griffiths

1. INTRODUCTION

The twenty-first century is not only an era of Research Infrastructures (RIs), but also an era of RI roadmaps. As discussed in the introduction to this volume, RIs are a varied category of facilities, resources and services that are used by scientific communities to conduct research (Cramer et al, ch 1 in this volume). RI roadmaps, as defined by the Global Science Forum (GSF) of the Organisation for Economic Co-operation and Development (OECD), are "strategic plans elaborated jointly by scientists and policymakers, under the aegis of the latter, with well-defined explicitly-stated contexts, goals, procedures and outcomes" (OECD 2008: 3). Similarly, the European Strategy Forum on Research Infrastructures (ESFRI) highlights in its first RI roadmap in 2006 that roadmaps "will allow policy makers to make better decisions in a wider European context in which national policies may be developed and possibly where it will be better to join a project at a pan-European level rather than try to develop an independent national facility" (ESFRI 2006b). While both establish RI roadmaps as tools to support decision making and strategic planning of policymakers, they emphasize two different areas of coordination, in the first case between scientists and policymakers, and in the second between the national and the European levels. Another specific feature of the ESFRI roadmap is that it includes very different types of RIs and therefore also aims to support coordination across different scientific fields. Today, most countries in Europe have introduced roadmapping exercises for RIs, which largely follow the model of ESFRI. And while the work of ESFRI certainly contributed to the increasing popularity of such RI roadmaps, its contribution to increased coordination across Europe is still limited. This is because, similarly to the case of RIs themselves, the category of national RI roadmaps in Europe is still very varied in scope and in terms of purposes (Bolliger et al 2017). Therefore, we argue that while RI roadmaps in Europe added a new layer in the narrative of RIs, highlighting the tensions between international collaboration and national interest, ESFRI has not (yet) achieved its intention to prioritize pan-European projects over national facilities.

This chapter outlines the developments behind this argument by investigating the context of current roadmapping processes at the national, European and international levels. Therefore, the chapter includes the historical description of the implementation and evolution of the ESFRI roadmapping processes, as well as a broad analysis of national roadmapping exercises. The purpose of the mapping of these different developments is to form a type of inventory to illustrate how and to what extent roadmapping as a strategic exercise has gained importance for agenda setting and prioritization in science policy in recent decades. The chapter focuses mainly on activities in Europe, including both national activities of European countries and the common European roadmapping activities through ESFRI.

Since the early 2000s, an increase in roadmapping activities for science policy and RI policy purposes can be observed. The general idea of these roadmaps can be traced back to technology roadmaps, which were mainly used within industry as tools or technique for supporting the management and planning of technologies (Phaal et al 2004). Since the 1970s, technology roadmapping has gained significant recognition within corporations as a tool to align technology strategy with business strategy (Cho et al 2016). The vision behind RI roadmaps owes much to this initial technological orientation in being, first and foremost, a tool to support strategic and long-term planning for companies to survive in increasingly competitive technological environments (Phaal et al 2004). The use of roadmaps in policymaking has gained in popularity (Carayannis et al 2016), and there is much to suggest that their use for planning scientific infrastructure projects outside Europe inspired the first ESFRI roadmap. In the European context, the preparation and publication of RI roadmaps has in effect become the main task of ESFRI since its formal launch in 2002 by the European Council as an expert group tied to the European Commission (Pilniok 2015), with the purpose to facilitate cooperation and funding of some of the European RIs (Ryan 2015). In the same period, several European countries such as Sweden, the Netherlands and France have started their own roadmapping processes and begun to prepare and publish national RI roadmaps (VR 2006; NWO 2008; MESRI 2008).

These roadmapping activities were originally driven by the creation of the European Research Area (ERA), through which the European Commission aimed to increase the competitivity of European science (Chou and Gornitzka 2014). Despite the great attention that RIs and RI roadmaps are receiving in several policy forums on the national, international and European levels, as briefly illustrated above, the topic has only received marginal interest in scholarly research. Previous investigations by the Horizon 2020 project InRoad, set up to contribute to the harmonization of policymaking around RIs at the European and national levels (InRoad 2018a), concluded that current roadmapping activities at the national level in Europe vary greatly in scope and purpose, and that there is also a lack of definition and shared understanding of what purpose roadmap exercises serve and through which process this purpose should be served (Bolliger et al 2017). The scope of RI roadmaps can vary from very broad descriptions of general objectives to more specific or sectorial planning, for example for different areas of science. Some RI roadmaps have an alleged focus on identifying gaps and needs in the existing national RI landscape (e.g. Sweden, United Kingdom (UK)); others also used it to guide the formulation of research policy with strategic RI priorities (e.g. Denmark, Norway, Israel). The division of roles between scientists and policymakers in those processes, as well as the content of those strategic plans, are equally diverse (Bolliger et al 2017).

This chapter builds upon previous work of the authors within the context of the Horizon 2020 project InRoad. The project investigated national RI roadmapping exercises in Europe with the aim to formulate recommendations towards a higher degree of coordination between national and European RI roadmapping processes (InRoad 2018a, 2018b; Bolliger et al 2017; Griffiths 2019). The analysis is furthermore based on the investigation of several policy documents that directly relate to the issue of roadmaps and RI and that include, for instance, roadmap documents, related strategy reports or governmental agendas, e.g. ESFRI documents, examples of RI roadmaps outside Europe, reports of the OECD GSF and national RI roadmaps.

The chapter is structured as follows. The section that follows this introduction describes the context behind the formation of ESFRI, including relevant initiatives at the international level. The detailed description of the creation of the first ESFRI roadmap is followed by an outline of the major changes in the evolution of ESFRI with regards to the following editions of the roadmap. The last section includes an overview and analysis of national RI roadmaps in Europe.

2. SETTING THE SCENE: THE CONTEXT OF THE CREATION OF ESFRI

The political interest in RIs and the introduction of this category in European research policy documents and strategy reports result from a shift in European policy towards an increased political integration of the issues of science and research, occurring as part of the launch of the ERA framework (Breschi and Cusmano 2004; Luukkonen et al 2006; Luukkonen and Nedeva 2010). This shift led to a new dynamic in European scientific collaboration towards the involvement of European Union (EU) and EU-related actors in matters that were previously self-organized by scientific communities, nationally focused and multilaterally resolved without EU involvement. Such examples include the European Space Agency, the European Molecular Biology Organization (EMBO), the Institut Laue-Langevin (ILL) and the European Synchrotron Radiation Facility (ESRF), which were all established through ad hoc collaborations between countries (Tindemans 2009; Guzzetti 1995; Cramer, ch 3 in this volume). While the "initial period of European organisation building [was] characterised by powerful, organised elites in few isolated research fields who exerted influence at the national level", according to Nedeva (2013: 224), this new dynamic also brought increasing coordination and governance of collaborative research in other scientific fields, including social sciences and humanities (Moskovko et al 2019; Hallonsten 2020; Franssen, ch 7 in this volume).

The ERA and the resulting RI policy activities at the European level were at the core of the Lisbon Strategy, launched in 2000 with the strategic goal "to become the most competitive and dynamic knowledge-based economy in the world" by the next decade (European Council 2000). While the success of the Lisbon Strategy and this particular goal can be questioned (Duchêne et al 2009), the ERA is still used as a framework for policymaking. In January 2000, the European Commission issued the memorandum *Towards a European Research Area*. The document included propositions for guidelines to support future research policy and initiated a broad discussion on research cooperation in Europe. Since the launch of ERA and the Lisbon Strategy, a new rationale for Europeanization has emerged, with the coordination of national budgets and ESFRI as examples of the resulting adoption of the Open Method of Coordination (Barré et al 2012; Chou and Gornitzka 2014; Flink 2016; Tamtik 2016). This was introduced by the European Commission in 2003 as a research policy initiative to foster mutual policy learning through benchmarking, action planning and exchange of best practices (Tamtik 2016). To achieve this, the Commission brings together state representatives to discuss different matters related to research policy, to exchange in an apparently non-binding way (Flink 2016). Specific examples, beyond ESFRI, include benchmarking exercises that compare national research systems (e.g. Research and Innovations Observatory reports), or the networking of national research programs (the so-called ERA-nets). According to Barré et al (2012), the Lisbon agenda provides a good illustration of this coordination concept. Furthermore, the Lisbon Strategy put knowledge at the center of the EU agenda, as a means to solve problems in areas such as economic growth, regional development, environmental issues and labor markets (Chou and Gornitzka 2014). According to Ulnicane (2015: 35), ERA represents a new generation of research governance, building upon previous ideas of innovation and economic thinking, but putting more emphasis on "collaboration, diffusion and the need for a systemic approach".

This development towards ERA includes longer historical trajectories in Europe. Since the early 1990s, in the context of the early European Framework Programmes for Research and Development, the stepwise creation of a common EU research policy has been recognized as important not only for competitiveness but also for ensuring greater cohesion within the EU and among its member states. In addition, the persistently lower level of investment into research and development across the EU at this time, compared to the United States (US) and Japan, gave rise to concerns that Europe might not successfully achieve the transition to a knowledge-based economy, considered to be the condition for global competitiveness. This was additionally worsened by the fragmentation, isolation and compartmentalization of national research systems as well as the inconsistency of regulatory and administrative frameworks (Breschi and Cusmano 2004). As mentioned in the introductory chapter (Cramer et al, ch 1 in this volume), this new generation of research governance represents a new understanding of the role of science for economic competitiveness in European science policy throughout the years (Flink and Kaldewey 2018).

In September 2000, a couple of months after the launch of the Lisbon Strategy and the publication of the ERA memorandum, the European Commission, together with the European Science Foundation and the French Ministry of Higher Education, Research and Innovation, organized a conference addressing different aspects of science policy related to RIs, and "provide a basis to design the broad lines of a European approach to research infrastructures, which could then be implemented jointly by the Member States and the EU in the context of the 'European research area'" (European Commission 2000). This conference was the first of an ongoing series of international conferences for RIs, which "have become a major information platform for stakeholders, policy makers and researchers to discuss issues concerning the realisation and operation of research infrastructures of pan-European and global relevance" (ESFRI 2009b: 16). During the first conference in 2000 in Strasbourg, a proposal was set up to form a permanent body to advise the European Commission on a coordinated approach to RIs across Europe. It was proposed that such a permanent body would "pave the way for scientific needs to become political funding decisions, in collaboration with the commission and EU member states" (Schiermeier 2000: 443).

In June 2001, the Research Council, which is a configuration of the Council of the EU attended by representatives from each EU member state at a ministerial level, invited the European Commission in collaboration with the EU member states and the EU associated and candidate countries "to study the best means of providing independent scientific advice as well as to explore new

Table 5.1 Overview of ESFRI members and their years of membership according to the ESFRI roadmaps

2002	Austria, Belgium, Denmark, Finland, France, Germany, Greece, Ireland, Italy, Luxembourg, Netherlands, Portugal, Spain, Sweden, United Kingdom
2003	Cyprus, Czech Republic, Estonia, Hungary, Latvia, Lithuania, Malta, Poland, Slovak Republic, Slovenia
2004	Bulgaria, Iceland, Israel, Liechtenstein, Norway, Romania, Switzerland, Turkey
2008	Albania, Croatia, Serbia
2010	Montenegro
2016	Macedonia, Moldova, Ukraine
2019	Armenia, Bosnia and Herzegovina, Iceland

arrangements to support policies on research infrastructures" (Dahmen 2001: 2). The resulting high-level Expert Group with representatives from all EU member states issued a report in February 2002, recommending the creation of a "European Strategy Forum on Research Infrastructures, ESFRI". This recommendation was followed by a letter from Philippe Busquin, member of the European Commission with responsibility for research, to the ministers of research (and similar) of the EU 15 member states (the countries of the EU prior to the 2004 enlargement), with the request to nominate representatives to the forum and offering operational support from the commission. ESFRI's first plenary meeting took place on 25 April 2002 in Brussels, with representatives from all EU member states and the European Commission (ESFRI 2006b). In November 2003, ESFRI was extended to the ten new EU member states that joined the EU during the enlargement round of 2004, and at the end of 2004, seven associated countries joined (see Table 5.1).

ESFRI is a self-regulated body composed of delegates from the EU member states and associated countries, with a secretariat provided by the European Commission. To date, the forum holds regular meetings, has an executive board and an elected chair who holds a non-renewable two-year mandate. In Table 5.1 all countries that became members of ESFRI during the last years are listed, based on information included in the ESFRI documents and on the ESFRI website (ESFRI 2019a).

Already prior to initiatives related to ERA, RIs had attracted increased policy attention on the international level. In 1992 the countries of the OECD established the Megascience Forum, with the mandate to encourage international cooperation and common policy strategies for large-scale RIs. To achieve this, the Megascience Forum formed a number of working groups, such as for neutron sources in 1993, and for radio astronomy in 1997. The creation of the Megascience Forum itself was a response to the trend "for ever-larger and more expensive instruments as well as an expression of the high hopes for greater international research cooperation after the fall of the Iron Curtain" (Kaiserfeld 2013: 30). In 1999 the forum was transformed into the GSF to consolidate and expand the work of its predecessor. The GSF membership includes 34 countries that are either OECD members or relevant partners, and the EU, which are represented by national delegates from academia, funding agencies or science ministries, during the bi-annual plenary meetings.

Since the launch of ESFRI in 2002 these two fora exchanged on their work. ESFRI consulted the GSF's work on future large-scale projects in astronomy and astrophysics, when it explored existing activities (ESFRI 2004). Shortly after the publication of the first ESFRI roadmap, the GSF put RI roadmapping on its agenda. Based on a list of strategic documents, which also included the first and second ESFRI roadmaps, the GSF organized an international workshop on "Enhancing the Utility and Policy Relevance of Roadmaps of Large Research Infrastructures", and on the basis of the results of the workshop published its "Report on Roadmapping of Large Research Infrastructures" (OECD 2008), which provided for the first time an analysis of the international "roadmapping landscape" at the time. Many of these roadmaps were sectorial and developed by scientific institutions and agencies, such as CERN, the US National Aeronautics and Space Administration and the European Space Agency. Cross-sectorial roadmaps were only found in Europe. Due to the lack of consensus on the definition of the terms "research infrastructures" and "roadmaps", the GSF adopted "roadmap" as "strategic plans elaborated jointly by scientists and policymakers, under the aegis of the latter, with well-defined explicitly-stated contexts, goals, procedures and outcomes" (OECD 2008: 3). These roadmaps were understood as strategic, long-range planning exercises to support decision-making processes. Their increasing popularity was explained with the need of policymakers to take into consideration the priorities and requirements of scientific communities, the international context and societal priorities when facing decisions about the planning, funding and implementation of large RIs. Furthermore, the report concluded that while the purposes of roadmaps vary greatly, they all reflect advances in policymaking processes, "beyond past practices in which proposals for large infrastructures were considered separately based on lobbying by strongly motivated individuals or communities of scientists" (OECD 2008: 3). The activity resulted in the 2010 publication of a further report on "Establishing Large International RIs: Issues and Options" (OECD 2010). In the following years, the GSF investigated other related topics, such as international distributed RIs (OECD 2014), sustainability and effectiveness of international RIs (OECD 2017) and the assessment of scientific and socio-economic impact of RIs (OECD 2019).

The increased attention of coordinating RI policies at the global scale was not limited to the activities of the OECD countries. In addition to the GSF, at its first science ministers meeting in Okinawa in 2008, the G8 countries decided to set up the ad hoc Group of Senior Officials (GSO) for "Global Research Infrastructures". The objective of this group was to develop a non-binding international framework for the establishment of global RIs. The GSO is composed of representatives from Australia, Brazil, Canada, China, the European Commission, France, Germany, India, Italy, Japan, Mexico, Russia, South Africa, the UK and the US. Participating countries are represented on the GSO by government officials and experts in the areas of international research facilities and international relations (GSO 2015).

3. THE FIRST ESFRI ROADMAPS: THE BIRTH OF THE RI ROADMAP

When ESFRI started its work in 2002, exploring options to fulfill its mandate, it took inspiration from international examples of strategic plans for RIs already existing at that time. The most prominent example of such a document includes the "Facilities for the Future of Science: A 20-Year Outlook", which was developed by the Office of Science of the US Department of Energy and included a prioritized list of 28 major facilities (Department of Energy 2003). The European Commission referred to this document as a "classic case of how priorities for large-scale research infrastructures should be set to ensure the growth of a country in terms of technology, innovation and knowledge" (European Commission 2005: 6). During the same period, also the US National Institutes of Health had elaborated a long-term plan to support prioritized RIs for biomedical research (Check 2003) and launched the Roadmap for Medical Research in September 2004 (National Institutes of Health 2019).

The US Department of Energy and National Institutes of Health were not alone in preparing such long-term strategy documents. In Australia, the National Collaborative Research Infrastructures Strategy was conceived in 2004 as a program that strategically invested in RI in a coordinated way across the nation and was part of the "Backing Australia's Ability – Building our Future through Science and Innovation" package, an initiative introduced by the Australian government in 2001 to support science and innovation (Nicoll 2005). It was a result of the work of a "National Research Infrastructure Task Force", which was similar to ESFRI, convened in 2003 "to develop a National Research Infrastructure Strategic Framework to inform Government investment in research infrastructure for universities and publicly funded research agencies" (Sargent 2004). The final report of the taskforce recommended work with roadmaps, and these were consequently published in 2006, 2008 and 2011 as National Collaborative Research Infrastructure Strategy (NCRIS) Strategic Roadmaps, and in 2016 under the new name National Research Infrastructure Roadmap. Since 2010 Australia and the EU have developed formalized collaboration in RI policy (NCRIS 2015). In the same period, the Japanese Ministry of Education, Culture, Sports, Science and Technology planned to concentrate its further investment on facilities with large multidisciplinary user communities, and in South Africa, the hosting of international RIs was also made a policy objective, seen as an indispensable tool for national growth (European Commission 2005). ESFRI invited representatives from all these countries to the third European Conference for Research Infrastructures, taking place in Nottingham in December 2005, to present their national strategic documents (ESFRI 2006a).

After two years of existence, during the Dutch presidency of the Council of the EU in November 2004, ESFRI was invited by the Competitiveness Council, the council configuration attended by the ministers from the member states and the European commissioners responsible for the area Internal Market, Industry, Research and Space, to prepare a roadmap for new large-scale RIs or substantial upgrades of existing RIs needed by the European scientific community, according to criteria that were still largely to be defined. The council stressed that the "roadmap should describe the scientific needs for research infrastructures for the next 10 to 20 years, on the basis of a methodology recognized by all stakeholders and take into account input from relevant intergovernmental research organisations as well as the industrial community" (Council of the European Union 2004).

To do so, ESFRI brought together policymakers and scientists, including scientists from different fields and experts from intergovernmental research organizations, such as the European International Research Organisation Forum (EIROforum), in an effort consisting of consultative work over nine months (ESFRI 2006a). Three international Roadmap Working Groups (RWGs; later renamed Strategy Working Groups) were established: Biological and Medical Sciences, Physical Sciences and Engineering and Social Sciences and Humanities. They were assisted by a number of experts, the majority of them from Europe but also from the rest of the world, including Russia and the US, that were selected on the basis of their scientific expertise and their contribution to science policy, and their international reputation (ESFRI 2006b). During the work on the first ESFRI roadmap, the forum investigated existing national strategic exercises and their respective methodologies to inform the development on the European level. Therefore, during the ESFRI meetings, different approaches and experiences from the national level were discussed, including examples from the UK, Italy, Germany, Denmark and Sweden, with the latter two having published documents on their national needs for further investment in RIs in December 2005 and October 2006, respectively (ESFRI 2006a).

During the ESFRI meeting of September 28–29, 2006 the forum agreed on a first list of 35 proposals for new or major upgrades of facilities of pan-European interest. The final version of the first ESFRI roadmap was published on October 19, 2006. The document stated that it was not a list of priorities, which was certainly a necessary statement, due to the lack of legitimacy of the European Commission, or ESFRI in this case, to define European funding priorities (Ryan 2015; Tamtik 2016). Instead the aim of the list was described as to "facilitate discussion and allow for coherent planning", as well as to "stimulate other groups to look at pan-European initiatives of a similar nature that might be added in future editions" (ESFRI 2006b: 15). Furthermore, the roadmap described that for projects to be included in its future editions, they had to meet the following three elements: being "of pan-European relevance (and) provide unique opportunities for world-class research" (ESFRI 2006b: 10), and applying "an 'Open Access' policy for basic research, i.e. be open to all interested researchers, based on open competition and selection of the proposals evaluated on their sole scientific excellence by international peer-review" (ESFRI 2006b: 16). The definition of RIs in the roadmap also acknowledges that RIs can be in any field, and that they can be "single-sited", "distributed" or "virtual" (ESFRI 2006b). After the publication of this first roadmap in 2006, ESFRI was invited by the Competitiveness Council to update the document on a regular basis. The 2006 roadmap was followed up with new iterations in 2008 and 2010, and in 2016 and 2018. Currently, the ESFRI roadmap 2021 is in preparation.

4. THE EVOLUTION OF ESFRI: THE GROWTH OF A NEW FORUM

ESFRI has undergone several changes with regard to its internal structure and organizational framework since its creation in 2006. In the process of the 2008 update, the RWGs have evolved with the identification of additional areas of importance for the EU research landscape such as environmental sciences and e-infrastructure. The RWG for Environment was established at the end of 2006 to identify new RIs of pan-European relevance for Energy, and Biological and Medical Sciences and Environmental Sciences (ESFRI 2008b). An e-Infrastructures Working Group was set up as a cross-disciplinary entity, including representatives from all four RWGs and from the e-Infrastructure Reflection Group (ESFRI 2007). Like ESFRI, the e-Infrastructure Reflection Group is a European self-regulated independent body of national delegates, founded in 2003 to provide strategic advice and guidance on the development of European e-infrastructures such as grid infrastructures, cloud computing and data repositories (e-IRG 2005). This stronger focus on e-infrastructures and distributed RIs resulted in a revision of the definition of RIs for the 2008 update of the ESFRI roadmap, namely that RIs could adopt a single-site or multiple-site structure according to their specific technical characteristics and mission (ESFRI 2008b).

In order to support the participation of the countries joining the EU in 2004 or later (sometimes called the EU 13 member states) in RI activities at the pan-European level, a specific ESFRI Working Group was devoted to regional issues. The RI-related conferences during the Slovenian EU presidency in 2008 and the Czech EU presidency in 2009 were focused on the structural and regional dimension of ERA. After the association of the Western Balkan states to the EU's Seventh Framework Programme in early 2008, invitations to nominate ESFRI delegates were extended to Albania, Croatia, Macedonia, Montenegro and Serbia (ESFRI 2008a).

In its first two roadmaps, ESFRI had put considerable effort into identifying scientific needs for new RI projects or substantial updates of existing RIs. For the following update, the forum committed to take a proactive role towards the realization of the RIs included in its roadmaps (ESFRI 2009b). This turn away from identifying additional RI projects for the roadmap and instead focusing on the already identified projects coincided with the economic crisis at the time, which put national research budgets under pressure. The 2010 ESFRI roadmap was published in early 2011, including six new RIs, and was incorporated into a wider strategy document, the ESFRI Strategy Report (ESFRI 2010).

In the period that followed, the ESFRI working groups underwent further changes. The RWGs were transformed into Strategy Working Groups (SWGs) with a stronger focus on so-called "Grand Challenges" and future policy-oriented needs. This shift of focus reflected the increasing attention given to the socio-economic impact of RIs in the European research policy of the time. Following a seminar on "the Role of Research Infrastructures for a Competitive Knowledge Economy", which was organized by the European Commission together with ESFRI (ESFRI 2009a), the resulting Lund Declaration from July 2009 emphasized "that European Research Policy should move away from the present bureaucratic structure and instead focus on the Grand Challenges to the World – e.g. climate change, water shortage, demography and pandemics" (ESFRI 2009a).

Another important change occurred during this period, through the rapidly increasing rate of assessment exercises for the RIs included in the roadmaps at the European and national levels (Duşa et al 2014). After the publication of its first three roadmaps (in 2006, 2008 and 2010), ESFRI got a mandate to focus on supporting the implementation of the projects on the roadmap to achieve the goal, set by the Innovation Union Flagship Initiative, to complete or launch the construction of 60 percent of the projects by 2015. Therefore, in 2011, the ESFRI Implementation Group was established to identify basic criteria for RIs considered to be under implementation (Duşa et al 2014) and in 2012 the Assessment Expert Group was appointed with the task to evaluate the financial and managerial maturity of the RI projects included on the ESFRI roadmap 2010 (Calvia-Goetz et al 2013). All these changes were reflected in the 2016 roadmap.

ESFRI's work with the 2016 roadmap update followed the Conclusions of the Council of the EU from May 2014, calling on ESFRI to continue the prioritization of all ESFRI projects (Council of the European Union 2014). To achieve this, the five SWGs made a first ever analysis of the European landscape of RIs, under the general criteria of international open access to facilities and data, a peer review-based selection of proposals and the excellence and uniqueness of scientific services provided. The results of this analysis were later included as a Landscape Analysis in the roadmap. Furthermore, the SWGs reviewed all the ten-year-old projects from the 2006 ESFRI roadmap to identify so-called ESFRI "landmarks". ESFRI landmarks refer to RI projects that have been successfully launched and therefore can be considered major factors for competitiveness of the ERA. The projects that had been included in the ESFRI roadmaps of 2006, 2008 or 2010 were reviewed by the Implementation Group, according to a pattern devised by the Assessment Expert Group in 2012. With regard to proposals for new projects or upgrades, the evaluation was done in parallel with the SWGs reviewing the scientific case, e.g. scientific merit, relevance and impact, European added value and the Implementation Group assessing the maturity of the proposals, e.g. stakeholder commitment, user strategy and access policy (ESFRI 2016).

For the latest ESFRI roadmap, of 2018, the Landscape Analysis from the ESFRI roadmap 2016 was updated and the ESFRI projects and landmarks were grouped in the six thematic domains: Energy, Environment, Health and Food, Physical Sciences and Engineering, Social and Cultural Innovation, and Data, Computing and Digital RIs (ESFRI 2018). Table 5.2 provides a summary of the mentioned changes and activities of ESFRI.

While ESFRI's work on its roadmaps was not the first exercise of this type, in a broader perspective, the ESFRI roadmap was the first international cross-border roadmap developed jointly by policymakers and scientists from across Europe and encompassing the following scientific domains: Social Sciences and Humanities; Environmental Sciences; Energy; Biomedical and Life Sciences; Materials Sciences; Astronomy, Astrophysics, Nuclear and Particle Physics; and Computation and Data Treatment. The publication of the first ESFRI roadmap set in motion notable changes and laid ground for a widened international RI policy discussion on different issues such as funding mechanisms, criteria for evaluation and monitoring of RIs, international collaboration and so on. Many countries started work on their own national RI roadmaps, and today almost all European countries have published at least one edition of a national RI roadmap or similar document. Similarly, research communities were encouraged to better coordinate and provide structured input into roadmapping cases, which led to more diverse RIs being

Table 5.2 Overview of ESFRI-related activities

2000	*First European Conference for Research Infrastructures 2000, Strasbourg*
2002	**Establishment of the European Strategy Forum on Research Infrastructures (ESFRI)**
2003	Inclusion of ten new EU member states *Second European Conference for Research Infrastructures 2003, Trieste*
2004	Inclusion of seven associated countries Set-up of three Roadmap Working Groups (RWG) –Physical Sciences and Engineering –Biological and Medical Sciences –Social Sciences and Humanities
2005	*Third European Conference for Research Infrastructures 2005, Nottingham* Updated ESFRI guidelines Set-up of 15 expert groups Revision of ESFRI charter and decision to create Executive Board
2006	***ESFRI Report 2006, including 35 ESFRI projects***
2007	*Fourth European Conference for Research Infrastructures 2007, Hamburg* New RWG Environment New ESFRI Working Group e-infrastructure Working Group on Regional Issues
2008	***ESFRI Roadmap 2008, including 44 ESFRI projects*** *Fifth European Conference for Research Infrastructures 2008, Paris* New ESFRI Working Group Regional Issues Slovenian Presidency Conference: Research Infrastructures and their structuring dimension within the ERA
2009	***ESFRI Roadmap Implementation Report 2009*** New RWG for Energy Czech Presidency Conference: Research Infrastructures and the regional dimension of ERA
2010	***ESFRI Strategy Report on Research Infrastructures. Roadmap 2010*** New ESFRI Working Group evaluation *Sixth European Conference for Research Infrastructures 2010, Barcelona* Belgian Presidency Conference: Infrastructures for Energy Research
2011	***ESFRI Evaluation Report 2011*** RWG became Strategic Working Groups –Energy –Health and Food –Environment –Social and Cultural Innovation –Innovation and Emerging Technology New ESFRI Working Group Implementation
2012	***ESFRI Roadmap Implementation Report 2012*** *International Conference for Research Infrastructures 2012, Copenhagen* Assessment Expert Group Report 2013

2013	*New ESFRI Working Group on Innovation*
2014	*International Conference for Research Infrastructures 2014, Athens*
2015	New ESFRI WG on Investment Strategies in e-Infrastructures
2016	***ESFRI Strategy Report on Research Infrastructures. Roadmap 2016*** *International Conference for Research Infrastructures 2016, Cape Town* New ESFRI Working Group on Long-Term Sustainability
2018	***ESFRI Roadmap and Strategy Report 2018*** *International Conference for Research Infrastructures 2018, Vienna* Bulgarian Presidency Conference: Research Infrastructures beyond 2020 – sustainable and effective ecosystem for science and society ESFRI Workshop: Monitoring of Research Infrastructures, periodic update of landmarks, use of key performance indicators
2019	ESFRI Workshop: The future of Research Infrastructures in the European Research Area
2020	*International Conference for Research Infrastructures 2020, Ottawa*
2021	***Forthcoming: ESFRI Roadmap 2021***

included in the ESFRI roadmaps and a rising number of policy activities for RIs in e.g. humanities (see Franssen, ch 7 in this volume). The 13 countries that joined the EU in 2004 and 2007 have also become more involved in RI policies and, corresponding to the ambition of ERA to coordinate national efforts, attributing to the integrative nature of European RI policy activities.

With regards to the evolution of the purpose of these roadmaps, Griffiths (2019) identified different phases in the development of ESFRI and its RI roadmapping process. The first phase was characterized by the incubation of projects. The first roadmap was developed through extensive consultations and broad landscape analyses, to foster the development of new pan-European RIs. In 2008, the constriction of research budgets caused by the financial crisis led the European Council to request more prioritization. In response, ESFRI worked towards including into the roadmapping process an assessment of the implementation of the projects and priority setting for RI projects. During this phase, a High-Level Expert Group was formed by the European Commission to evaluate the implementation status of projects on the ESFRI roadmap. This resulted in the introduction of new, stricter guidelines for ESFRI projects. The results of this prioritization phase were actualized in the process for the 2016 roadmap, which also introduced a more systemic approach to RI roadmapping – the "ecosystem approach" (ESFRI 2016). The "ecosystem" is understood as the ensemble of diverse RIs currently active, emerging or being phased out in Europe, and the ambition is to minimize redundancies and increase efficiency within this system. To achieve this, ESFRI attempts to analyze the overall landscape of RIs, and especially complementarity between projects. This phase is ongoing, with an increasing focus on assessing the implementation of RIs. Finally, the next phase will be defined by potential ways forward, which will likely be defined by challenges and priorities. Some have to do with ESFRI's internal structure and processes, such as evaluation and monitoring methodologies and ensuring that they are fit for purpose. An additional challenge is for ESFRI to find a role in the global context, and there are more comprehensive attempts to link science with economic competitiveness, as well as "Grand Challenges" such as climate change (Griffiths 2019).

5. THE RISING POPULARITY OF NATIONAL RESEARCH INFRASTRUCTURE ROADMAPS: THE OFFSPRING

In the following section the focus lies on national RI roadmaps in Europe. A brief overview of the evolution of the European RI roadmap landscape as well as a look at the current processes behind these roadmaps shall help to better understand the influence ESFRI had on the coordination of RI policies in Europe. The implementation of national RI roadmaps is a priority for ESFRI (ESFRI 2019b) and was welcomed by its chairman, John Wood, in the foreword of the first ESFRI roadmap in 2006:

> The role of ESFRI in fostering incubation and stimulation will be exercised to bring as many of these projects to a point where decisions by ministers are possible. This requires in the first place discussions and decisions at national level in particular as regards the lead role one country or several countries may wish to take for certain projects. These reflections might encourage the development of national roadmaps and the earmarking of dedicated national budgets for the construction of Research Infrastructures with a European/international dimension, which ESFRI would welcome. (ESFRI 2006b: 7)

Therefore, for the launch of the first ESFRI roadmap in 2006, John Wood, chair of ESFRI at the time, wrote personal letters to each minister for research (and equivalent) in the EU member states and associated countries to encourage discussions at the national level and to trigger reflections on the development of national roadmaps and the earmarking of dedicated national budgets for the construction of European RIs (ESFRI 2006a). Since the publication of ESFRI's first roadmap, national authorities were repeatedly invited to take stock of their capacities and needs of national communities to start these prioritization exercises. During the ESFRI forum meetings, national delegates were asked to report on developments in their countries. This strong interest in such prioritization exercises on the national level was largely due to the fact that the funding necessary to support the implementation of the ESFRI projects needed to be secured on the national level (ESFRI 2009a). Therefore, ESFRI monitored national activities, and each of the ESFRI annual reports included an overview of the status of national roadmapping exercises. Currently, according to the monitoring of national roadmaps on the ESFRI website, all EU member states have national roadmaps in place, with the exception of Iceland, Latvia, Luxembourg, Malta and Slovakia. Additionally, the following ESFRI associated countries also have introduced roadmaps: Montenegro, Norway and Switzerland. In the case of Belgium, Cyprus and Turkey, it is indicated that roadmaps are in preparation (ESFRI 2019b).

Recent results of the EU-funded project InRoad revealed that these roadmaps are very diverse in scope and purpose, and that timing and methodologies behind their elaboration vary (Bolliger et al 2017). Moreover, the project's analysis confirmed further differences with regard to the governance of the national roadmap processes, funding mechanisms for RIs and evaluation and monitoring methodologies (InRoad 2018b). The InRoad project included several work packages dedicated to collecting data on those processes to identify good practices and inform policymakers. While the final recommendations of the project will not be considered in this chapter, as the project pursued the specific goal of enhancing coordination of RI roadmapping in Europe, it provided a wealth of data on national RI roadmap processes in Europe. Unprecedented so far, and therefore of particular interest, is the empirical data, which were collected through a broad online consultation, addressing the authorities responsible for the national RI roadmaps (Bolliger et al 2017), as well as a number of expert interviews. These semi-structured expert interviews were executed by the authors of this chapter with different actors involved in national RI roadmap processes in a number of countries (InRoad 2018a; Griffiths 2019). These data enable an assessment of the influence of the ESFRI roadmap on national levels.

Before taking a closer look at current national roadmapping processes, an overview of national RI roadmaps in Europe is apt. Table 5.3 shows an overview of the existing national RI roadmaps in Europe, based on an extensive desk review completing the information on national RI roadmaps, available on the website of ESFRI (2019b). Documents that have been included are those which refer to the national level and were produced by the entity responsible for the funding of RIs in the respective countries, which can include contributions to RI of national relevance as well as national contributions to international collaborations. Roadmaps which do not cover the entire national landscape were not considered. For this reason, examples such as the Helmholz-Roadmap for RIs in Germany from 2011, which only deals with RIs within the Helmholz Association, or the different roadmaps of single research councils in the UK, preceding their consolidation into UK Research and Innovation, were not included. The same accounts for national roadmaps for a specific type of RI such as the Swedish "Outlook for the National Roadmap for e-infrastructures for Research" from 2019.

Table 5.3 Overview of existing national RI roadmaps in Europe

EU 15 countries	
Denmark	– Danish Roadmap for Research Infrastructures 2011
	– Danish Roadmap for Research Infrastructures 2015
Finland	– National-Level Research Infrastructures – Present State and Roadmap (2009)
	– National-Level Research Infrastructures – Present State and Roadmap 2014–2020
	– National-Level Research Infrastructures – Present State and Roadmap 2014–2020. Interim Review Report 2018
France	– Research Infrastructures for France – Roadmap 2008
	– Research Infrastructures – Roadmap 2012–2020
	– French National Strategy on Research Infrastructures – 2016 Edition
Germany	– Roadmap for Research Infrastructures: A Pilot Project of the Federal Ministry of Education and Research (2013)
Greece	– Greek Large-Scale Research Infrastructures Roadmap: A 10-Year Outlook (2007)
	– National Roadmap for Research Infrastructures 2014
Ireland	– Research Infrastructure in Ireland – Building for Tomorrow 2007
Italy	– Roadmap Italiana delle Infrastrutture di Ricerca di Interesse Pan-Europeo 2011
	– Programma Nazionale per le Infrastrutture di Ricerca 2015–2020
Netherlands	– The Netherlands' Roadmap for Large-Scale Research Facilities 2008
	– The Netherlands' Roadmap for Large-Scale Research Facilities 2012
	– National Roadmap Large-Scale Scientific Infrastructures 2016
Portugal	– Portuguese Roadmap of Research Infrastructures 2014–2020

Spain	– Spanish Map of Unique Scientific and Technical Infrastructures (2007) – Spanish Map of Unique Scientific and Technical Infrastructures. Update (2013)
Sweden	– The Swedish Research Council's Guide to Infrastructures. Summary 2006 – The Swedish Research Council's Guide to Infrastructures – Recommendations on Long-Term Research Infrastructures by the Research Councils and VINNOVA. 2007 – The Swedish Research Council's Guide to Infrastructures. 2012 – The Swedish Research Council's Guide to Infrastructures. 2014 – The Swedish Research Council's Guide to Infrastructures. 2018
United Kingdom	– The UK's Research and Innovation Infrastructure: Opportunities to Grow Our Capability (2019)
EU 13 countries	
Bulgaria	– Republic of Bulgaria Resolution No. 692. September 2010. For Acceptance of National Roadmap for Research Infrastructure – Bulgarian National Roadmap for Research Infrastructure 2017–2023
Croatia	– Croatian Research and Innovation Infrastructures Roadmap 2014–2020
Czech Republic	– Roadmap for Large Research, Development and Innovation Infrastructures in the Czech Republic 2010 – Roadmap for Large Research, Development and Innovation Infrastructures in the Czech Republic May 2011 Update – Roadmap of Large Infrastructures for Research, Experimental Development and Innovation of the Czech Republic for the Years 2016–2022 – Roadmap of Large Infrastructures for Research, Experimental Development and Innovation of the Czech Republic for the Years 2016–2022. Update 2019

Estonia	– Estonian Research Infrastructures Roadmap 2010 – Estonian Research Infrastructures Roadmap 2014 – Estonian Research Infrastructures Roadmap 2019
Lithuania	– Roadmap for Research Infrastructures of Lithuania 2011 – Lithuanian Roadmap for Research Infrastructures 2015
Poland	– Polska Mapa Drogowa Infrastruktury Badawczej (2011) – Polska Mapa Drogowa Infrastruktury Badawczej (2014)
Romania	– Report regarding Research Infrastructures of Romania (2007) – Romanian Roadmap of Research Infrastructures 2017
Slovenia	– Research Infrastructure Roadmap 2011–2020 – Research Infrastructure Roadmap 2011–2020. Revision 2016
Associated countries	
Montenegro	– Montenegrin Research Infrastructure Roadmap 2015–2020 – Revised Roadmap for Research Infrastructure of Montenegro (2019–2020)
Norway	– Tools for Research – Part I: Norway's National Strategy for Research Infrastructure 2012–2017 – Tools for Research – Part II: Norwegian Roadmap for Research Infrastructure 2016
Switzerland	– Schweizer Roadmap für Forschungsinfrastrukturen Schlussbericht 2011 – Swiss Roadmap for Research Infrastructures in View of the ERI Dispatch 2017–2020 (Roadmap for Research Infrastructures 2015) – Swiss Roadmap for Research Infrastructures in View of the ERI Dispatch 2021–2024 (Roadmap for Research Infrastructures 2019)

The list of documents in Table 5.3 provides a first overview of the national RI roadmaps in Europe, which itself provides a number of insights. A number of countries introduced a national roadmap shortly after the publication of the first ESFRI roadmap in 2006, such as Greece, Ireland, Romania and Spain (in 2007). Sweden published its first roadmap in 2006 and updated it in 2007. In the year of the second ESFRI roadmap (2008), France and the Netherlands introduced their roadmaps, with Finland following in 2009. With the publication of the ESFRI roadmap update in 2010, national RI roadmaps were introduced in a larger group of countries, most of which belonged to the new (2004 and 2007) member states, with Bulgaria, the Czech Republic and Estonia in 2010, Lithuania and Poland in 2011 and Slovenia in 2012. Additionally, roadmaps were introduced in Italy and Denmark in 2011, and the EU associated countries Switzerland in 2011 and Norway in 2012. In Germany and Croatia, the first and so far only national RI roadmaps were introduced in 2013 and 2014, respectively. The German document includes the specification that the roadmap was a pilot project of the Federal Ministry of Education and Research (BMBF 2013). Among the countries where the national RI roadmap was introduced the earliest, Ireland is the only case where no update has followed to date. The latest roadmap on the list is the British one, which was published in 2019. Since a large number of these national RI roadmaps indicate a timeframe until 2020, it will be interesting to see how these national RI roadmaps develop in the coming years.

Also, the number of documents per country reveals large differences with regards to the frequency of the update of the roadmap. In a number of countries, regular cycles can be identified which are likely to coincide with their national financial planning cycles, such as in Sweden, Switzerland and Estonia, where these cycles last four years. In the case of Sweden and the Czech Republic, the first roadmap was updated after one year, and after a break, a regular cycle of updates was introduced. This can be explained by the novelty of the process and the need to adjust, e.g. due to internal pressures from the research community as in the Czech case (Griffiths 2019). In both cases, the initial roadmap process was adapted following analyses of relevant governance systems. In the Czech case, the Ministry of Education, Youth and Sport had mandated an international audit team with one of the most comprehensive studies of the national research system (Arnold 2011), and in the Swedish case, the former vice-chancellor of Stockholm University Kåre Bremer investigated the organization, governance and financing of Swedish national RIs on behalf of the Swedish Research Council (Bremer 2013).

The titles of the listed documents in Table 5.3 reveal that only in the case of Italy and Poland has no roadmap yet been published in English, which shows that the audience is mainly national and that coordinating national priorities with the European level may not be a priority. The majority of the documents include the terms "Research Infrastructures" and "roadmap", but there are exceptions: The Dutch documents refer to either "research facilities" or "science infrastructures", the Swedish documents are "Guides to Infrastructure" and the Spanish are called "Maps of Unique Scientific and Technical Infrastructures". The Spanish case indicates a desire to broaden the term, which is similar to the Croatian document's use of the term "Research and Innovation Infrastructures", and even more obvious in the Czech document, which speaks of "Research, Development and Innovation Infrastructures" (2010, 2011) and "Infrastructures for Research, Experimental Development and Innovation" (2016, 2019). Changes that occurred in the titles of the other national roadmaps show rather a convergence towards the use of "roadmap" and "Research Infrastructures".

The empirical data collected by the InRoad project allow us to gain a better understanding of the processes behind these documents. The results of the consultation carried out by the project showed that RI roadmapping processes usually combine different mechanisms and instruments, which are related to their purpose. In a majority of the reported cases, the RI roadmapping process has been viewed as an input for funding decisions (Bolliger et al 2017; Griffiths 2019).

Other purposes relevant to the majority of roadmap processes are the identification of strategic priorities foreseen for funding, the identification of scientific needs and existing gaps and to guide priority setting in research policy with strategic RI priorities (Bolliger et al 2017). The number of countries where the national roadmap is explicitly linked to the ESFRI roadmap and negotiations at the international level is limited. Overall, the purposes of national RI roadmapping exercises are far from unified across Europe (Griffiths 2019). This indicates limitations in the integrative effect of ESFRI's efforts on RI policies at the national level.

With regard to the design of the processes behind the national RI roadmaps, there are more overlaps, as more than two thirds include the following elements: (bottom-up) calls for applications, prioritization of RIs or projects, scientific evaluation of RIs, monitoring of projects and RIs included in the roadmap and support from independent national experts. Nonetheless, considerable variation between countries exists in the way these elements are implemented and used within the roadmapping process, and publicly available information on how these processes are precisely carried out is often lacking (Griffiths 2019). In the majority of cases, the scientific excellence of RIs and new projects must be established by an international expert panel before they can be included onto the roadmap. However, the selection processes for experts as well as the methodologies and criteria applied in the evaluations vary considerably across countries. Nonetheless, the criteria of uniqueness of an RI (at national level) and open access to the RI are common to all investigated roadmap processes and also reflect ESFRI's definition of RIs.

As already discussed above, the timelines for updating national roadmaps vary. In many roadmaps there is no explicit reference to the periodicity of updates, and further investigations revealed that there are many constraints around carrying out regular updates, especially in smaller countries where resources are relatively limited. Moreover, the process itself can be subject to changes in response to developments in RI policies and EU funding instruments, or as a result of systemic changes at the national level (Griffiths 2019). For example, the consolidation of the UK Research Councils into UK Research and Innovation (UKRI) facilitated the elaboration of the first national RI roadmap, encompassing all nine research councils.

6. DISCUSSION AND CONCLUDING REMARKS

The analysis in this chapter shows that the roadmap category for RIs was inspired by sectorial roadmaps with a focus on one scientific field such as astrophysics or energy, and as it has been done in the US, Europe and Australia. While different agencies, such as the US Department of Energy, the US National Aeronautics and Space Administration (NASA) and the European Space Agency (ESA) elaborated long-term plans for the facilities needed to conduct specific types of research in their respective scientific field, the OECD GSF was instrumental in defining and conceptualizing these plans as RI roadmaps. By the time that this was made, ESFRI had already been established and started working on its first roadmap, which was published in 2006, and in parallel, Sweden had developed the first national roadmap for RIs in Europe. The ESFRI roadmap, which is inspired by those previous exercises, is a novelty in its form as it is an international exercise and encompasses a wide array of scientific fields. Furthermore, it is unique since it is published by a forum without any formal authority or a formal budget, but with the "hidden" aim of encouraging an alignment of national funding priorities along a list of pan-European RIs. The "Open Method of Coordination", for which the ESFRI roadmap process is exemplary, and to which the InRoad project was supposed to contribute through benchmarking and exchange of best practices, has its limitations in view of securing the long-term funding of such large-scale endeavors as pan-European RIs. This is because national policymakers have a considerable amount of leeway when it comes to how they design and implement their national RI roadmaps.

The main difference between ESFRI and the national RI roadmaps is the funding responsibility: There is no budget at the European level to finance the projects and landmarks on the ESFRI roadmap. The considerable burden of funding RIs is shouldered by individual countries in Europe, with the exception of some "preparatory phase" funding provided within Horizon 2020. Therefore, the success of the ESFRI process relies on national authorities setting in place mechanisms to support the identification, prioritization, construction and implementation of pan-European RIs. Even at the national level, these processes are necessarily quite complex, due to the considerable costs of such infrastructures, the combination of technical and political goals as well as the diversity of national systems in Europe. The fulfilment of the ambitious aim behind ESFRI, to encourage the prioritization of joining projects at the pan-European level over developing independent national facilities in national policies (ESFRI 2006b), therefore, not only requires more time but also considerably more political support, particularly by the countries which do not rely on access to pan-European RIs as they are able to fund their own internationally competitive RIs.

While a certain integrative effect can be attributed to ESFRI, considering the heavily expanded membership, and the inclusion of broader scientific fields through additional ESFRI Working Groups, it is limited in the respect of securing funding needed to implement the ESFRI priorities in the national budgets. This has much to do with difficulties of coordinating priorities already at the national level, but further in-depth investigation at the national level, covering different cases, is needed.

Although this chapter provides only a first glimpse into the topic and leaves out several important aspects, such as funding or governance of the RIs on the ESFRI roadmap, it provides an outline of the relevant dynamics and illustrates the complexity of analyzing RI roadmaps. This chapter therefore indicates a large potential for further research. For example, a more in-depth focus at the national level would allow a better understanding of the effects of the EU's "Open Method of Coordination" on national RI policies, and how these coordination processes are influenced by national systems. It would also be interesting to investigate more recent initiatives at the European level, such as the "European Open Science Cloud", and the role of ESFRI, as well as national processes within it.

ACKNOWLEDGMENTS

Part of the chapter is based on the contributions of both authors to the InRoad project, which was funded by the EU's Horizon 2020 Research and Innovation program under grant agreement No 730928.

REFERENCES

Arnold E (2011) *International audit of research, development and innovation in the Czech Republic*. Technopolis Group.

Barré R, L Henriques, D Pontikakis and M Weber (2013) Measuring the integration and coordination dynamics of the European Research Area. *Science and Public Policy* **40** (2): 187–205.

BMBF (2013) *Roadmap for research infrastructures: A pilot project of the Federal Ministry of Education and Research (BMBF)*. Available at https://ec.europa.eu/research/infrastructures/pdf/roadmaps/germany_national_roadmap_en.pdf (last accessed December 15, 2019).

Bolliger I et al (2017) *InRoad consultation report: Prioritisation, evaluation and funding of Research Infrastructures in Europe*. Available at http://inroad.eu/wp-content/uploads/2017/11/InRoad_Consultation_Report_201711.pdf (last accessed December 11, 2019).

Bremer K (2013) *Synpunkter på planering, organisation, styrning och finansiering av svensk nationell infrastruktur*. Available at www.vr.se/download/18.2412c5311624176023d25bb8/1529480527943/Synpunkter-planering-org-styrning-finansiering-sv-infrastruktur_VR_2013.pdf (last accessed December 15, 2019).

Breschi S and L Cusmano (2004) Unveiling the texture of a European Research Area: Emergence of oligarchic networks under EU Framework Programmes. *International Journal of Technology Management* **27** (8): 747–72.

Calvia-Goetz A et al (2013) Assessing the projects on the ESFRI roadmap: A high level expert group report. Available at https://ec.europa.eu/research/evaluations/pdf/archive/other_reports_studies_and_documents/esfri.pdf (last accessed January 17, 2020).

Carayannis, E, A Grebeniuk and D Meissner (2016) Smart roadmapping for STI policy. *Technological Forecasting and Social Change* **110**: 109–16.

Check E (2003) NIH "roadmap" charts course to tackle big research issues. *Nature* **425** (438). Available at www.nature.com/articles/425438b#citeas (last accessed January 13, 2020).

Cho Y, S Yoon and K Kim (2016) An industrial technology roadmap for supporting public R&D planning. *Technological Forecasting and Social Change* **107**: 1–12.

Chou M and Å Gornitzka (2014) *Building a European knowledge area: An introduction to the dynamics of policy domains on the rise*. Edward Elgar Publishing.

Council of the European Union (2004) Press release. *2624th Council Meeting Competitiveness* (Internal Market, Industry and Research), Brussels, November 25–26. Available at https://ec.europa.eu/commission/presscorner/detail/en/PRES_04_323 (last accessed December 11, 2019).

Council of the European Union (2014) *Council conclusions on the implementation of the roadmap for the European Strategy Forum on Research Infrastructures*. Available at http://data.consilium.europa.eu/doc/document/ST-10257-2014-INIT/en/pdf (last accessed January 17, 2020).

Dahmen A (2001) MEMO/01/245. Results Research Council, June 26. Available at https://ec.europa.eu/commission/presscorner/detail/en/MEMO_01_245 (last accessed December 18, 2019).

Department of Energy (2003) *Facilities for the future of science: A 20-year outlook*. Available at https://fribusers.org/frib/docs/2003-DOE20YearScienceFacilities.pdf (last accessed December 11, 2019).

Duchêne V, E Lykogianni and A Verbeek (2009) The EU R&D under-investment: Patterns in R&D expenditure and financing. In H Delanghe, U Mudur and L Soete (eds) *European science and technology policy: Towards integration or fragmentation*. Edward Elgar Publishing.

Duşa A, D Nelle, G Stock and G Wagner (eds) (2014) *Facing the future: European research infrastructures for the humanities and social sciences*. Scivero.
e-IRG (2005) e-Infrastructures opportunity list. Available at http://e-irg.eu/documents/10920/12353/Opportunities+list+2005 (last accessed December 11, 2019).
ESFRI (2004) *European Strategy Forum on Research Infrastructures: Report 2004*.
ESFRI (2006a) *Annual report 2005–2006*. Available at www.esfri.eu/annual-reports (last accessed December 11, 2019).
ESFRI (2006b) *European Roadmap for Research Infrastructures: Report 2006*. Available at www.esfri.eu/sites/default/files/esfri_roadmap_2006_en.pdf (last accessed December 11, 2019).
ESFRI (2007) *Annual report 2006–2007*. Available at www.esfri.eu/annual-reports (last accessed December 11, 2019).
ESFRI (2008a) *Annual report 2008*. Available at www.esfri.eu/annual-reports (last accessed December 11, 2019).
ESFRI (2008b) *European Roadmap for Research Infrastructures: Roadmap 2008*. Available at www.esfri.eu/sites/default/files/esfri_roadmap_update_2008.pdf (last accessed December 11, 2019).
ESFRI (2009a) *Annual report 2009*. Available at www.esfri.eu/annual-reports (last accessed December 11, 2019).
ESFRI (2009b) *European Roadmap for Research Infrastructures: Implementation report 2009*. Available at www.esfri.eu/sites/default/files/esfri_implementation_report_final2_hidef.pdf (last accessed December 11, 2019).
ESFRI (2010) *Strategy report on Research Infrastructures: Roadmap 2010*. Available at www.esfri.eu/sites/default/files/esfri-strategy_report_and_roadmap_2010.pdf (last accessed December 11, 2019).
ESFRI (2016) *Strategy report on Research Infrastructures: ESFRI Roadmap 2016*. Available at www.esfri.eu/roadmap-2016 (last accessed December 11, 2019).
ESFRI (2018) *Strategy report on Research Infrastructures: ESFRI Roadmap 2018*. Available at http://roadmap2018.esfri.eu/media/1066/esfri-roadmap-2018.pdf (last accessed December 11, 2019).
ESFRI (2019a) *ESFRI delegates*. Available at www.esfri.eu/people/delegates (last accessed December 18, 2019).
ESFRI (2019b) *The national Roadmaps*. Available at www.esfri.eu/national-roadmaps (last accessed December 18, 2019).
European Commission (2000) *Infrastructures: The backbone of European research*. A major conference in Strasbourg (F), September 18–20. Press Release. Available at https://ec.europa.eu/research/press/2000/pr1209-infras-en.html (last accessed December 11, 2019).
European Commission (2005) *Third European Conference on Research Infrastructures, Nottingham, UK 6 and 7 December 2005, main conclusions*. Available at https://ec.europa.eu/research/evaluations/pdf/archive/fp6-evidence-base/evaluation_studies_and_reports/evaluation_studies_and_reports_2005/european_conference_on_research_infrastructures_december_2005.pdf (last accessed December 11, 2019).
European Council (2000) Lisbon European Council March 23–24, *Presidency conclusions*. Available at www.europarl.europa.eu/summits/lis1_en.htm (last accessed January 17, 2020).
Flink T (2016) EU-Forschungspolitik – von der Industrieförderung zu einer pan-europäischen Wissenschaftspolitik? In D Simon et al (eds) *Handbuch Wissenschaftspolitik*. Springer, pp 79–97.

Flink T and D Kaldewey (2018) The new production of legitimacy: STI policy discourses beyond the contract metaphor. *Research Policy* **47**: 14–22.

Griffiths A (2019) *Analysing research infrastructure roadmapping processes: The case of the European Strategy Forum for Research Infrastructures and its influence on national processes*. Master thesis, University of Lausanne.

GSO (2015) *Group of senior officials on global Research Infrastructures progress report 2015*. Available at https://ec.europa.eu/info/sites/info/files/research_and_innovation/gso_progress_report_2015_final.pdf (last accessed December 11, 2019).

Guzzeti L (1995) *A brief history of European Union research policy*. Office for Official Publications of the European Communities.

Hallonsten O (2020) Research infrastructures in Europe: The hype and the field. *European Review* **28** (4): 617–35.

InRoad Consortium (2018a) *Good practices and common trends of national research infrastructure procedures and evaluation mechanisms (Deliverable D3.3)*. Available at http://inroad.eu/reports/ (last accessed December 11, 2019).

InRoad Consortium (2018b) *InRoad final report*. Available at http://inroad.eu/reports/ (last accessed December 11, 2019).

Kaiserfeld T (2013) The ESS from neutron gap to global strategy. In T Kaiserfeld and T O'Dell (eds) *Legitimizing ESS: Big Science as a collaboration across boundaries*. Nordic Academic Press.

Luukkonen T and M Nedeva (2010) Towards understanding integration in research and research policy. *Research Policy* **39**, 674–86.

Luukkonen T, M Nedeva and R Barré (2006) Understanding the dynamics of networks of excellence. *Science and Public Policy* **33** (4), 239–52.

MESRI (2008) *Research Infrastructures for France. Roadmap 2008*. Available at https://cache.media.enseignementsup-recherche.gouv.fr/file/Infrastructures_de_recherche/62/7/roadmap_TGIR_2008_527627.pdf (last accessed December 11, 2019).

Moskovko M, A Astvaldsson and O Hallonsten (2019) Who is ERIC? The politics and jurisprudence of a governance tool for collaborative European Research Infrastructures. *Journal of Contemporary European Research* **15** (3): 249–68.

National Institutes of Health (2019) *NIH Roadmap and Roadmap-affiliated initiatives*. Available at www.niehs.nih.gov/funding/grants/announcements/roadmap/index.cfm (last accessed December 17, 2019).

NCRIS (2015) *Australia's relationship with the European Union on research infrastructure*. Available at www.education.gov.au/australia-s-relationship-european-union-research-infrastructure (last accessed December 18, 2019).

Nedeva M (2013) Between the global and the national: Organising European science. *Research Policy* **42** (1), 220–30.

Nicoll C (2005) *National Collaborative Research Infrastructure Strategy (NCRIS)*. Presentation at the Third European Conference for Research Infrastructures, December 6–7, Nottingham.

NWO (2008) *The Netherlands' Roadmap for large-scale research facilities*.

OECD (2008) *Report of roadmapping of large research infrastructures*. Available at www.oecd.org/sti/inno/47057832.pdf (last accessed December 11, 2019).

OECD (2010) *Large Research Infrastructures*. Available at www.oecd.org/sti/inno/47057832.pdf (last accessed December 11, 2019).

OECD (2014) *International distributed Research Infrastructures: Issues and options*. Available at www.oecd.org/sti/inno/international-distributed-research-infrastructures.pdf (last accessed December 11, 2019).

OECD (2017) *Strengthening the sustainability and effectiveness of international research infrastructures*. Available at www.oecd-ilibrary.org/docserver/fa11a0e0-en.pdf?expires=1575891582&id=id&accname=guest&checksum=49C05E28D6B2B36CD16EAC413BBE2E26 (last accessed December 11, 2019).

OECD (2019) *Reference framework for assessing the scientific and socio-economic impact of research infrastructures*. Available at www.oecd-ilibrary.org/docserver/3ffee43b-en.pdf?expires=1575891646&id=id&accname=guest&checksum=0E714B52DAC85AFFFC0B7E2D4D700AEE (last accessed December 11, 2019).

Phaal R, C Farrukh and D Probert (2004) Technology roadmapping: A planning framework for evolution and revolution. *Technological Forecasting and Social Change* **71**: 5–26.

Pilniok A (2015) The institutionalisation of the European Research Area: The emergence of transnational research governance and its consequences. In A Franzmann, A Jansen and P Münte (eds) *Legitimizing science: National and global publics (1800–2010)*. Campus Verlag, pp 275–306.

Ryan L (2015) Governance of EU research policy: Charting forms of scientific democracy in the European Research Area. *Science and Public Policy* **42**: 300–14.

Sargent M (2004) National Research Infrastructure Framework: The final report of the National Research Infrastructure Taskforce. Available at https://docs.education.gov.au/system/files/doc/other/national_research_infrastructure_taskforce_final_report_2004.pdf (last accessed December 18, 2019).

Schiermeier Q (2000) Europe urged to set up advisory body on research infrastructure. *Nature* **407**: 433–4.

Tamtik M (2016) Institutional change through policy learning: The case of the European Commission and research policy. *Review of Policy Research* **33** (1): 5–21.

Tindemans P (2009) Post-war research, education and innovation policy-making in Europe. In H Delanghe, U Mudur and L Soete (eds) *European science and technology policy: Towards integration or fragmentation*. Edward Elgar Publishing.

Ulnicane I (2015) Broadening aims and building support in science, technology and innovation policy: The case of the European Research Area. *Journal of Contemporary European Research* **11** (1): 31–49.

VR (2006) *The Swedish Research Council's guide to infrastructures: Summary 2006.*

6. Intensified role of the European Union? European Research Infrastructure Consortium as a legal framework for contemporary multinational research collaboration

Maria Moskovko

1. INTRODUCTION

What may such diverse scientific installations as a particle accelerator, a network of a thousand ocean observation platforms and a collection of mice DNA models have in common? Clearly, all of them are scientific resources that enable or support work of various communities of research users. They are collaborative by their nature, since an effort of a single institution or a country is rarely enough to establish and operate such resources. Despite the different missions they serve and forms they undertake, the above-mentioned projects are all Research Infrastructures (RIs). In the context of the European Union (EU), the overarching term of RI refers to "facilities, resources and related services that are used by the scientific community to conduct top-level research in their respective fields" (European Council 2009). The examples of the diverse scientific installations mentioned above demonstrate a variety of multinational RIs that were set up and operate under a legal-administrative arrangement called the European Research Infrastructure Consortium (ERIC). It was introduced by the EU as a policy instrument specifically designed to facilitate the set-up and operation of pan-European collaborative RIs.

European cooperation on various types of Big Science and RIs (Hallonsten 2014; Krige 2003; Papon 2004) has traditionally been taking place without any direct involvement of the EU and its predecessors since the early 1950s. One exception is the International Thermonuclear Experimental Reactor (ITER) in which the European Atomic Energy Community, represented by the European Commission, became a founding member in 2006 together with China, India, Japan, Russia, South Korea and the United States. Other major European research organizations have to this day been set up as intergovernmental organizations under treaty agreements (e.g. the European Organization for Nuclear Research (CERN), the European Molecular Biology Laboratory (EMBL) and the European Southern Observatory (ESO)) or established as limited liability companies under the national law of the host countries (the European Synchrotron Radiation Facility (ESRF) and Institut Laue-Langevin (ILL) are French limited liability companies; the European X-Ray Free-Electron Laser (XFEL) is a German limited liability company) (Cramer, ch 3 in this volume).

The enactment of ERIC by the EU in 2009 meant the creation of a new organizational structure specifically for science, in order to enable the set-up and operation of multinational RIs, based on international agreements, in which countries and international organizations (IOs) can be members; ERICs are, hence, public-public partnerships to their core. The ERIC status grants RI a legal personality and recognizes it as a European Community (EC) body (Preamble 6, European Council 2009) and an international body exempt from value added tax (VAT), excise duties and procurement laws of the EU (European Council 2009, Preamble 10, Articles 5(d), 7(3)).

As per late 2019, 21 RIs have been established with ERIC status, and more are envisioned in the near future (ESFRI 2018). Nineteen of those RIs have also been included into one or several of the European Strategy Forum on Research Infrastructures (ESFRI) roadmaps (Bolliger and Griffiths, ch 5 in this volume). Indeed, ERICs are by definition envisioned as "European Research Infrastructures at Community level which are necessary for the efficient execution of the Community's RTD [Research and Technological Development] programmes" (European Council 2009, Article 5), however, the ERIC regulation does not necessitate their inclusion in the ESFRI roadmap.

Scholars of different fields have started to devote attention to specific issues pertaining to the ERIC form: intellectual property rights (Yu et al 2017; Ryan 2019), taxation (Hilling et al 2017), procurement (Graber-Soudry 2019) and the origins and initial steps of implementation of the policy instrument (Moskovko et al 2019). At first sight, the enactment of the ERIC form by the EU seems to signal an intensified level of EU involvement in science policy, however, the role of the EU has been marginal in the matter of RI policy so far, as demonstrated by the decades-long intergovernmental agreements among countries on constructing and operating collaborative research facilities (Hallonsten 2014; Krige 2003; Papon 2004). This chapter sets out to understand the role of the EU in the enactment and implementation of the ERIC instrument by focusing on the following three components: (1) how the novel legal-administrative form emerged on the EU policy agenda, (2) what characterizes organizations established with ERIC status and (3) how the policy tool has been applied in practice over the course of the past ten years.

The method of process tracing (George and Bennett 2005: 205ff) will be utilized in order to reconstruct the timeline and determine the actions and motives of the EU for enacting the ERIC legislation. It will be necessary to turn to EU law for understanding the legal basis upon which the EU acts and limitations and freedoms it may have in certain policy areas. Organizational characteristics of ERICs will be presented in a comparative perspective among the cases of individual ERICs, with process tracing utilized for determining the patterns of their creation. The chapter utilizes documents produced by different EU institutions and RIs established as ERICs. The author's attendance at three ERIC network meetings between 2016 and 2018 provided an opportunity for participant observation and interaction with representatives of the RIs and the European Commission and allowed to inquire on the first-hand experiences with the ERIC form. The notes produced at those meetings complement the list of material used for this work.

The chapter will start by explaining the legislative nature and the policy aspirations of the EU that made it possible to create the ERIC form. It will then present an overview of the established ERICs and culminate with the highlights of the implementation of EU Regulation 723/2009 (also referred to as the "ERIC Regulation"). The legal and administrative fundamentals of ERIC are summarized in Box 6.1. The concluding remarks will reflect on the nature of this policy instrument and the future that it may bring.

BOX 6.1 ERIC IN A NUTSHELL

ERIC (EC 723/2009) is a policy tool that grants RIs legal personality and allocates certain benefits to them similar to those of IOs, namely:

- relief from value added tax (VAT) and excise duties;
- exemption from the EU procurement laws;
- ability to craft own internal policies, including the one on public procurement.

It therefore provides a novel organizational form designed especially for enabling the set-up and operation of RIs. It is a unique practice in itself, considering that there has never been a legal-administrative form specifically for scientific organizations.

2. ERIC IS BORN

Despite the process of European political and economic integration taking place since the 1950s, intergovernmental collaboration on research and on the European Economic Community (EEC) were "evolving on two different tracks" (Papon 2009: 26). Namely, the EU has been criticized for lacking clearly pronounced policy frameworks (Papon 2004) and utilizing minor steering and financial mechanisms in the RI domain (Krige 2003). The reason for the "lack" of EU involvement has been the fact that despite the development and expansion of the policy portfolio of the supranational body with each of the treaties that led to the present-day EU, research has remained a domain of national priorities and budget allocations (Buonanno and Nugent 2013: 5–11).

The European Coal and Steel Community treaty of 1951, the European Atomic Energy Community treaty and the treaty establishing the EEC, both of 1957, implied research activity, however an explicit reference to research policy came only with the Single European Act (SEA) (1986, Article 24). The SEA put forward coordination and cooperation as methods of EU research policy and defined instruments, including multiannual funding programs, for policy implementation (SEA 1986, Article 24). Finally, the Treaty of Lisbon amending the Treaty on European Union and the Treaty establishing the EC (Treaty of Lisbon 2007) allocated the "shared competence" status to the areas of research, technological development and space (Treaty of Lisbon 2007, Article 2C (3)). The EU's competences in this area are thus exercised via the principle of subsidiarity. This principle strives to safeguard the ability of the member states to make decisions and take action, authorizing intervention by the EU in case "the objectives of an action cannot be sufficiently achieved by the Member States, but can be better achieved at Union level, by reason of the scale and effects of the proposed action" (European Parliament 2016). The principle of subsidiarity is often referred to as a mechanism for balancing power between individual member states and the integration objectives of the EU (Barton 2014; Öberg 2016).

Despite being a domain of national priorities and budget allocations, cross-border collaboration on science has been encouraged by various financial mechanisms from the side of the EU (Delanghe et al 2009). An even greater boost to EU involvement in the research and innovation domain came out from the Lisbon Strategy in 2000 devised with an ambition to "make Europe, by 2010, the most competitive and the most dynamic knowledge-based economy in the world" and hence prioritizing innovation, learning economy, as well as social and environmental revival (European Parliament 2000). The ambitious European Research Area (ERA) initiative of 2000 may be seen as one of the practical instruments of the strategy, focused on an increased and better coordinated European cooperation on science and technology initiatives, linked to attaining a "fifth freedom" – "movement of knowledge" (Chou 2014; Ulnicane, ch 4 in this volume).

RIs became one of the pillars of the ERA (Ryan 2015). As a result, in 2002 ESFRI was set up as a policy coordinating platform by the European Competitiveness Council, in order to streamline an EU-wide policy on existing and upcoming RIs (Papon 2009: 40). ESFRI is composed of delegates from the member states, associated countries and observers and a representative of the European Commission. The ESFRI launched a so-called roadmapping process for RIs, with the first roadmap published in 2006. Various installations were included in the list, not just tangible and single-sited facilities, but also resources for biomedical and life sciences, environmental sciences, as well as social sciences and humanities (ESFRI 2006).

Under the Lisbon Strategy, the EU funding for research entitled "Framework Programmes [FP] for Research and Technological Development" extended the seventh consecutive program (2007–13) for the seven-year funding period and €1.8 billion was allocated specifically for RI-related activities (European Parliament 2006). The FP 8 (2014–20), titled Horizon 2020, brought €2.5 billion for RI initiatives (European Parliament 2013). The proposal for the forthcoming Horizon Europe (2021–27) funding scheme budgeted the same amount of €2.5 billion for RIs (European Commission 2019g). Altogether, this signifies a remarkable increase from the initial 30 million European Currency Unit (an electronic unit of account that preceded the euro) that was allocated under the FP 2 (1987–91) for the "use of major installations" (European Council 1987).

The exponential budgetary increase of EU funding, specifically targeted at the "major installations" (wording used in the 1980s and 1990s) and "research infrastructure" (a word combination that emerged in the 2000s), signifies the prioritization of the matter and its alignment with the EU ambitions and goals. Interestingly, only somewhere around the turn of the millennium did "research infrastructure" gain an increased attention in FP 5 (1998–2002) and thus became pronounced as a fixed collocation of "RI" in the EU discourse (European Parliament 1999). Prior to that, FP 4 (1994–98) mentioned "large installations and large instruments", as well as the "information and communications infrastructure" (European Parliament 1994) and FP 3 (1990–94) referred to "scientific and technical infrastructure" (European Council 1990).

Infrastructure per se implies something tangible and concrete with a mission to provide a support function for other elements. Possibly, in the era of contested research funding, "infrastructure" also stipulated a longer-term outlook for something that needed to be sustained over the years while serving multiple user communities and hence implying a need for longer-term funding to ensure its thriving. In that sense, RI constitutes a boundary object (Star and Griesemer

1989; Guston 2001), as it may mean different things to different audiences, without alternating its form. While it ensures continuation of the academic work for scientific audiences, and implies further technological development of research-related resources and instruments, it may also be perceived as a sustainable object for the funders in view of its dissemination of knowledge into the society at large through the education of the younger generations and provision of industrial applications. The EU, in its turn, may be perceiving the support of RI as a way to mobilize the attainment of its ambitious policy goals tied to the knowledge economy and competitiveness on the global arena.

By this virtue European RIs received a label of "a pillar" of the Europe 2020 Strategy and were pictured as "engines" that were expected to drive forward the EU economy by advancing its science and technology (European Commission 2008c; ESFRI 2008). It should be mentioned that there were promoters of the matter of RI among various audiences: at the EU, at the member states, as well as on the level of individuals involved in different collaborative scientific projects. The initial reference to RI may be traced to September 2000, when the first conference on research infrastructures was held in Strasbourg under the French presidency in the European Council (ECRI 2006). The European Conference on Research Infrastructures (ECRI) brought forward a proposal to create a permanent advisory body to the EC specialized on the matter of scientific infrastructure (Schiermeier 2000), that in 2002 would become the ESFRI. The ECRI would take on a biannual tradition and in 2012 it changed its name to the International Conference on Research Infrastructure (ICRI).

Additionally, other initiatives on RI materialized with financial support from the European Commission via the FP funding scheme: Realising and Managing International Research Infrastructures (RAMIRI), running 2008–10 and later extended until 2013 (European Commission 2019a, 2019b), RICH 2020, the abbreviation for the European Network of National Contact Points for RIs in Horizon 2020, an EU-funded project to facilitate the network of National Contact Points and to promote and support RI-related matters, including application for RI-related funding under the Horizon 2020 program (European Commission 2019c). The problems of structural barriers around the set-up and implementation of RI were raised in the above-mentioned fora and at the ESFRI. Following the two workshops held in 2006 on the available legal forms that could accommodate RI-specific administrative and managerial issues, a proposal was made that in view of the upcoming pan-European RIs being distributed, a novel legal form may need to be created for their realization, "positioned between national law and the status of an international organisation [...] might help to solve many of the problems identified" (European Commission 2008a). In 2008 that proposal would materialize into the EC Proposal for a Council Regulation on the Community legal framework for a European Research Infrastructure (European Commission 2008b).

The 2009 enactment of ERIC without doubt signifies the importance of the matter of RI on the EU policy arena. RI in itself possesses characteristics of something that implies collaboration across boundaries and knowledge sharing that in their turn are also expected to ultimately contribute to producing socio-economic benefits such as novel scientific findings and innovations coming out from prospective industrial applications. All of the above-mentioned features may also be aggregated to the EU visions and goals of "unity in diversity", growth, cohesion, mobility, diplomacy, competitiveness on the global arena, etc. It may be posited that as much as some of the RI projects rely on the EU funding mechanisms and since 2009 also on its legal-administrative form (the ERIC form), the EU – viewed as a collective body, as well as the European Commission as its executive branch – to an extent also depends on the features of RI that align with the ambitious goals.

While DG Research and Innovation was the main addressee of the problems and possible solutions, the EU as a collective body had to come up with possible available solutions, in this case referred to as "spillovers", from one policy area that may be an inspiration of the other (Buonanno and Nugent 2013: 24). The existing legal-administrative forms of the European Economic Interest Grouping (EEIG) (European Council 1985) and the European Company (European Council 2001) could have served as an inspiration for the desired all-encompassing tool for RIs under the EU (see also OECD 2014). However, the unlimited liability of the EEIG and the economic and profit orientation of both the EEIG and the European Company forms were not considered plausible solutions (ECRI 2007). At the time of the ERIC making its way onto the EU political agenda, another novel instrument for cross-border cooperation of the EU members was in the making – the European Grouping of Territorial Cooperation (EGTC) (European Parliament 2006; Zapletal 2010). This however was more of service to the cohesion policy following the 2004 enlargement of the EU to Central and Eastern Europe rather than the research policy of the EU.

The enactment of the ERIC is a demonstration of an unprecedented action from the side of the EU in utilization of its legislative powers in order to create a legal form that enables the set-up and operation of collaborative RI projects. In accordance with Article 187 of the Treaty on the Functioning of the EU (TFEU 2012), the supranational body has the power to establish partnerships via so-called "joint undertakings" or "any other structure necessary for the efficient execution of Union research, technological development and demonstration programmes". Therefore, the EU made full use of Article 187 and created another "structure" – ERIC – intended to serve the needs of the contemporary collaborative RI projects.

Being a creation of the EU legislative powers, a balance needed to be struck with the degree of further EU involvement into the ERIC. The Galileo satellite navigation system, set up as a joint undertaking with the EU as a partner in 2002 (European Council 2002), was not considered a suitable option for becoming an ERIC, due to the EU's funding commitments to the project. Thus, EU involvement in the ERIC became explicitly limited and the participating countries were to be the funders: ERIC "should not be conceived as a Community body [...] but as a legal entity of which the Community is not necessarily a member and to which it does not make financial contributions" (European Council 2009). Moreover, the initially proposed name for this legal instrument – European Research Infrastructure (ERI) (European Commission 2008a) – ended up as the European Research Infrastructure *Consortium* (ERIC) – an indication of an intentional shift of responsibilities (possibly implying funding, as well) from the EU to the participating countries.

The EU however took on the role of facilitator, with the European Commission as an executive branch overseeing the process for application to the status of ERIC and streamlining the implementation of the ERIC regulation via the comitology procedure (European Council 2009, Article 20). The principle of subsidiarity also allows the EU to intervene in matters "only if and in so far as the objectives of the proposed action cannot be sufficiently achieved by the Member States [themselves], but can rather, by reason of the scale or effects of the proposed action, be better achieved at Union level" (Treaty on European Union 2012, Article 5). This demonstrates that the EU is not willing to be a full standing member (or funder) on par with the participating countries. The EU rather chooses to act within the limits of the competences, conferred upon by its members (Reichel et al 2014: 1055).

As has been pointed out above, the research policy domain is an area of shared competencies between the EU and its member states. While the legislative powers of the EU were fully utilized in order to create a legal framework that is capable of accommodating the prerequisites for the set-up and operation of multinational RIs, no further involvement of the EU is intended for realizing this legal-administrative form. From the policy perspective ERIC may be seen as a legal instrument enacted specifically with the purpose of overcoming structural barriers and simplifying and stimulating the establishment and operation of RIs (Moskovko et al 2019). The role of the EU, undertaken by the European Commission in the process of setting up ERICs, may be characterized by that of a facilitator, while the member states pursue "business as usual" in terms of multilateral negotiations on the statutory seat of the RI, monetary and in-kind commitments, agreements on access, and other matters. The enactment of ERIC demonstrates an increased role of the EU in the RI policy domain, however, it also demonstrates the EU's firm intention to remain an observer and a facilitator, rather than an equal stakeholder alongside the member states.

3. CHARACTERISTICS OF THE ESTABLISHED ERICS

When looking at the list of the 21 ERIC facilities established to date (December 2019) (Table 6.1), one can see that only one – the European Spallation Source – is a "tangible" research facility in the sense of a "Big Science" laboratory (Cramer et al, ch 1 in this volume). The remaining 20 are distributed networks and resources of various types of scientific instruments and their components.

This characteristic refers to the reflections over Big Science and RIs posed earlier in this volume (Cramer et al, ch 1 in this volume), leading into a further inquiry on what may have stimulated the growth of the present-day distributed RIs. To be more specific, what mobilized the loose networks of institutes, collaborating research groups and complementary scientific resources to establish their activity under a more tangible and durable structure of an RI, and particularly under the organizational form of ERIC?

Indeed, in order to apply for the status of ERIC, an entity needs to carry out "European research programmes and projects", contribute with an "added value in the strengthening and structuring of the ERA", provide "effective access" to those resources and facilities and ensure "mobility of knowledge" and "dissemination" (European Council 2009, Article 4). Those features are as broad as the nature of RI itself (Cramer et al, ch 1 in this volume), allowing various types of collaborative resources to get unified under the ERIC framework. A notion of a common identity within the cohort of the ERICs has been reinforced through some symbols – for example, the European Commission awards an identification plate at the initiation ceremony of a new ERIC, moreover a title of "ERIC" needs to be indicated next to the full name of the RI (European Council 2009, Article 8(2)).

When it comes to characteristics immanent to the established ERICs, most of them may be distinguished by provision of support functions and/or bringing together existing (predominantly distributed) resources and instruments that operate as national research infrastructure in different countries. Many of the organizations appear to have been collaborations among their current partners prior to obtaining the legal status of ERIC. A number of them were initiated as bottom-up initiatives from the scientific communities and supported through the EU FP funding schemes, for example, the Survey of Health, Ageing and Retirement in Europe (SHARE), the first ERIC established in 2011 (European Commission 2011). The SHARE initiative was consecutively funded by FP 5 (1998–2002), 6 (2002–2006) and 7 (2007–13) (SHARE 2016). The increased collaboration that eventually led to the set-up of the following ERIC, however, mostly came during the increased funding for RI under the FP 7 scheme. The so-called "preparatory phase" funding budget was allocated specifically to

Table 6.1 The 21 ERICs that exist to date (December 2019), in chronological order of ERIC status granted

Acronym	Full name	Form	ESFRI area	Statutory seat (country)	ERIC status granted	Capital value (M€)	Operation costs (M€)	No. of found. member states*	Funded under FP?
SHARE	The Survey of Health, Ageing and Retirement in Europe	Survey	Social and Cultural Innovation	Munich (DE)	2011	250	18	5	FP 5, 6, 7, 8
CLARIN	Common Language Resources and Technology Infrastructure	Data repository	Social and Cultural Innovation	Utrecht (NL)	2012	n/a	14	9**	FP 7
EATRIS	European Infrastructure for Translational Medicine	Support of translational research	Health and Food	Amsterdam (NL)	2013	500	2.5	8	FP 7
ECRIN	European Clinical Research Infrastructure Network	Support services to clinical trials	Health and Food	London (UK)	2013	5	5	5	FP 6, 7, 8
ESS	European Social Survey	Survey	Social and Cultural Innovation	Graz (AT)	2013	n/a	2.5	15	FP 5

Acronym	Full name	Form	ESFRI area	Statutory seat (country)	ERIC status granted	Capital value (M€)	Operation costs (M€)	No. of found. member states*	Funded under FP?
BBMRI	Biobanking and Biomolecular Resources Research Infrastructure	Support service to biomedical research	Health and Food	Paris (FR)	2013	195	3.5	16	FP 7
EURO-ARGO	EURO-ARGO	Support service to ocean observation	Environment	Brest (FR)	2014	10	8	9	FP 7
CERIC***	Central European Research Infrastructure Consortium	Network of national labs	Physical Sciences and Engineering	Trieste (IT)	2014	100	10	6	–
DARIAH	Digital Research Infrastructure for the Arts and Humanities	Digitalization service in arts and humanities	Social and Cultural Innovation	Paris (FR)	2014	4.3	0.7	15	FP 7
JIVE***	Joint Institute for VLBI ERIC	Support to VLBI astronomy	Physical Sciences and Engineering	Dwingeloo (NL)	2014 (transfer from a Dutch foundation)	n/a	2.5	4	FP 4, 5, 6, 7, 8

Acronym	Full name	Form	ESFRI area	Statutory seat (country)	ERIC status granted	Capital value (M€)	Operation costs (M€)	No. of found. member states*	Funded under FP?
–	European Spallation Source	Neutron spallation source	Physical Sciences and Engineering	Lund (SE)	2015 (transfer from a Swedish company)	1843	140	15	FP 7
ICOS	Integrated Carbon Observation System	Support of research from observation station	Environment	Helsinki (FI)	2016	116	24.2	9	FP 7
EMSO	European Multidisciplinary Seafloor and Water Column Observatory	Support of ocean observatory research	Environment	Rome (IT)	2016	100	20	8	FP 7
–	Life Watch	Support of biodiversity and ecosystem research	Environment	Seville (ES)	2017	150	12	8	FP 7

Acronym	Full name	Form	ESFRI area	Statutory seat (country)	ERIC status granted	Capital value (M€)	Operation costs (M€)	No. of found. member states*	Funded under FP?
ECCSEL	The European Carbon Dioxide Capture and Storage Laboratory Infrastructure	Support of carbon capture facilities	Energy	Trondheim (NO)	2017	1000	0.85	5	FP 7, 8
CESSDA	Consortium of European Social Science Data Archives	Hub for archives	Social and Cultural Innovation	Bergen (NO)	2017	117	39	15	FP 7
INSTRUCT	Integrated Structural Biology Infrastructure	Support of structural biology research	Health and Food	Oxford (UK)	2017	400	30	14**	FP 7
EMBRC	European Marine Biology Resource Center	Coordination and support of marine research	Environment	Paris (FR)	2018	164.4	11.2	9	FP 7
EU-OPEN-SCREEN	European Infrastructure of Open Screening Platforms for Chemical Biology	Support of chemical biology research	Health and Food	Berlin (DE)	2018	82.3	1.2	7	FP 7

Acronym	Full name	Form	ESFRI area	Statutory seat (country)	ERIC status granted	Capital value (M€)	Operation costs (M€)	No. of found. member states*	Funded under FP?
EPOS	European Plate Observing System	Support for solid Earth science	Environment	Rome (IT)	2018	32	18	12	FP 7
Euro-BioImaging	European Research Infrastructure for Imaging Technologies in Biological and Biomedical Sciences	Support of biological and medical imaging technologies	Health and Food	Turku (FI)	2019	90	1.6	15	FP 7

Note: * Including observers; **including international organizations; *** not on ESFRI roadmap.
Source: ESFRI (2016, 2018); EC decisions on ERICs; websites of RIs.

incubate future RIs that had been prioritized in the ESFRI roadmap process (ESFRI 2016, 2018). There is therefore an indication of a certain "push" from the ESFRI process towards the emergence of new RI initiatives and allocation of the "start-up" FP funding for their realization.

Indeed, the areas of activities of the ERICs are diverse and include the "strategic areas" put forward by the ESFRI – physical sciences and engineering, environment, energy, social and cultural innovation, and health and food (ESFRI 2016). Nineteen of the current ERICs were prioritized in the ESFRI process and only two – the Central European Research Infrastructure Consortium (CERIC ERIC) within materials science and the Joint Institute for Very Long Baseline Interferometry (JIVE ERIC), within astronomical interferometry for radio telescopes – were never prioritized by the ESFRI. However, as noted above, inclusion in the ESFRI roadmap is not a requirement of the ERIC regulation (European Council 2009, Article 4). Nonetheless, the majority of the ERICs appear in one way or another to be the "products" of the ESFRI process and the "preparatory phase" funding scheme available to initiatives prioritized by the ESFRI.

When looking back, an interesting observation can be made from the consequence of the establishment of ERICs (Table 6.1). The first two RIs to attain ERIC status were e-infrastructures in social sciences and humanities: SHARE ERIC and Common Language Resources and Technology Infrastructure (CLARIN) ERIC, established respectively in 2011 and 2012. In 2013 distributed RIs in the area of medical sciences joined the list: European Infrastructure for Translational Medicine (EATRIS) ERIC , European Clinical Research Infrastructure Network (ECRIN) ERIC and Biobanking and Biomolecular Resources Research Infrastructure (BBMRI) ERIC. Afterwards a number of RIs in the area of "Environment" were granted ERIC status: EURO-ARGO in 2014, followed by Integrated Carbon Observation System (ICOS ERIC) and European Multidisciplinary Seafloor and Water Column Observatory (EMSO ERIC) in 2016 and Life Watch in 2017 (see Table 6.1).

Remarkably, RIs in the area of physical sciences – a "traditional" domain of the typical Big Science research facilities (Cramer et al, ch 1 in this volume) – came at a later stage, with CERIC ERIC and JIVE ERIC in 2014, as well as the European Spallation Source ERIC in 2015. Notwithstanding the prioritization of the areas of energy and food by ESFRI, the first ERIC in the "Energy" domain – the European Carbon Dioxide Capture and Storage Laboratory Infrastructure (ECCSEL ERIC) – was established in 2017 in Norway. So far, no ERIC has been established in the "Food" domain although ESFRI prioritizes this category alongside "Health". The 2018 update to the ESFRI roadmap however pointed out that a new RI called METROFOOD – an RI promoting metrology in food and nutrition – was now a priority and that following the preparatory phase it is expected to get set up as an ERIC with a statutory seat in Italy (ESFRI 2018).

The established ERIC facilities are also diverse in the range of costs of their initial investments and operation. For example, the Digital Research Infrastructure for the Arts and Humanities (DARIAH ERIC) cost the least out of all ERICs: a mere €4.3 million to establish a budgeted €600 000 per year to operate (ESFRI 2016, 2018). The most expensive ERIC facility is the neutron science facility European Spallation Source, with projected construction costs of €1.8 billion and €140 million per year in operation costs (ESFRI 2016, 2018).

Some of the existing ERICs were constructed with the ERIC organizational form in mind from the very beginning (e.g. CERIC ERIC), while some switched from other legal-administrative forms when a need arose (e.g. European Spallation Source when it moved on to the construction phase in 2015, or the Cherenkov Telescope Array, that is currently undergoing a review of its application for ERIC and awaits the construction phase) (Moskovko and Teich forthcoming). Hence, a certain notion of "plasticity" of this form comes from the vague definition of what RI is and what it has to accomplish, along with a relatively easy-to-utilize framework that may be applied at any phase of the project's life cycle. However, one may need to ponder over what the limits of RI and ERIC may be. For example, can a Maltese initiative on block chain and machine learning for music qualify for an RI, and consequently an ERIC – the status that it intends to obtain (MusicNOW 2018)?

The locations of the established ERICs also point out a certain pattern. In 2016 about a quarter of all ERIC facilities were located in France and another quarter in the Netherlands. The first ERIC, SHARE ERIC, was initially established in the Netherlands in 2011 and three years after moved its statutory seat to Germany (European Commission 2011). The reason was apparently the constraints in the German national law to adopt the ERIC Regulation (BMBF 2019). CERIC is currently the only ERIC present in the Central and Eastern Europe region, stressing an asymmetry between the European "core" and "periphery" when it comes to the distribution of RIs and ERICs. Moreover, CERIC reportedly followed a top-down establishment via an implementing decision by the European Commission to launch the ERIC (European Commission 2014a) rather than a bottom-up initiation from the scientific communities. Several other RIs that are currently being established in Central and Eastern Europe are also in the process of applying for the ERIC status (the Extreme Light Infrastructure (ELI) and the International Centre for Advanced Studies on River-Sea Systems (DANUBIUS-RI)) and may soon balance out the geographic distribution of ERICs in Central and Eastern Europe (ICRI 2018). To what extent such a "balance" may be policy- rather than science-driven may be another question worth pondering upon.

The "plasticity" of ERIC mentioned above is also evident from the involvement of the associated countries of the EU in its implementation following the 2014 amendment to the ERIC Regulation, allowing the associated countries to become hosts to ERIC facilities (European Council 2013). An associated country of the EU is defined as a "third country which is party to an international agreement with the Community, under the terms or on the basis of which it makes a financial contribution to all or part of the Community research, technological development and demonstration programmers" (European Council 2009, Article 2(1) (c)). The agreement on the European Economic Area constitutes such an agreement, with Norway, Iceland and Liechtenstein able to become associated countries to ERIC. Hence, following the amendment, Norway became a host to ECCSEL ERIC and the Consortium of European Social Science Data Archives (CESSDA ERIC) in 2017. Norway had been involved in the latter RI since the 1970s, however on a basis of a memorandum of understanding collaboration and later a Norwegian limited liability company, prior to becoming an ERIC.

ERIC therefore appears to be a popular way of establishing collaborative RIs and demonstrates a certain degree of "plasticity" by being a relatively easy-to-use framework for the set-up and operation of RI provided that the founding member states commit to its funding. In practice it also means that the "rules of the game" remain the same for crafting the governance of ERIC as of any other RI: financial responsibilities of the shareholders, mitigation of rights and obligations through suitable governance arrangements, as well as outlining the benefits of "what is in it for me?" for each of the collaborating member states.

Upon their application for ERIC status, RIs are to draft their statutes – a constitutional document – in which they present the "essential elements" in form of their governance, liability matters, as well as the establishment of its statutory seat (European Council 2009, Article 10). Moreover, the RIs under ERIC are to draft their own internal policies that would serve as guiding principles for their day-to-day operations. Those may include data, access, intellectual property rights, employment policy, as well as procurement (European Council 2009, Article 10 (g)). Even though ERICs are exempt from the EU regulation on public procurement, they are still to follow the non-discrimination, transparency and competition principles; those values however may clash in the interplay between international public law when it comes to recognition of ERIC as an IO by the member states and the EU law that created ERIC (Graber-Soudry 2019).

The presented features of the established ERICs signal the diffuse characteristic of RI and confirm the "plasticity" of research infrastructure as a concept and as a policy domain. The process of mobilization of the collaborative European RIs to a certain extent gets orchestrated through processes established at the EU level: While ESFRI gives a go-ahead signal, the EU FP "preparatory phase" funding allows to "incubate" a prospective RI (ESFRI 2016, 2018). Such a scheme may be enough to give a head start to distributed initiatives that do not require vast investments. However, the set-up and operation of large-scale and tangible resources that rely on technologically advanced construction (e.g. the European Spallation Source and the upcoming Cherenkov Telescope Array and ELI) may be serving as a test for the envisioned intentions and actual possibilities that the EU legal form may or may not be able to provide.

4. IMPLEMENTATION AND EVOLUTION OF ERIC

A decade after the enactment of the ERIC legal-administrative form and the 21 organizations established with its help since then, an apparent need arose for these novel creations under EU law to hold regular meetings under the auspices of the so-called informal ERIC network to which the European Commission acts as a facilitator (European Commission 2014b, 2018). The intention of such gatherings is to "exchange best practices and raise common issues" among the established and prospective ERICs, and with the delegates from the ERIC Comitology Committee. Representatives of the ERIC Committee and of prospective ERICs that are in the final process of the ERIC application are invited to these meetings (European Commission 2014b). What appears to have started as the European Commission following through with its own tasks on ERIC (consultation meetings, comitology, collection of information for reports to EU institutions and member states), a certain degree of formation of a common identity among the established and prospective ERICs also started to take place. More so, the informal ERIC network (with meetings held between 2015 and 2018) is currently being shaped into a more formal structure – the ERIC Forum. The forum appears to act similarly to the EIRO Forum – a forum of major European intergovernmental research organizations, for example as demonstrated by the ability of the ERIC Forum to speak in one voice on a policy matter before the EU (ERIC Forum 2019).

Other initiatives for socialization among the existing and emerging ERICs may be witnessed through the FP-funded initiatives for the ERICs. Among those is a scheme for staff exchanges (European Commission 2019d) and a €1.5 million support for "more formal and structural collaboration and coordination of ERICs" (European Commission 2019e). The kick-off meeting for the latter initiative took place in January 2019 and a website along with social media accounts and hashtags were set up to ensure visibility of the forum. Further developments on socialization and institutionalization of ERIC will be interesting to follow, particularly in view of the increasing number of ERICs and the diverse areas, missions and user groups that they serve and the power dynamics the formalizing ERIC Forum may be exercising on the research policy arenas in the future.

When tracing the process of implementation of the policy tool of ERIC through the Commission-issued reports (European Commission 2014b, 2018) and through communication with representatives of the ERIC facilities, several tensions become evident: the undefined legal space, the novelty of the legal-administrative form, as well as unresolved structural problems that despite good intentions, ERIC has not yet been able to solve. The discussion below will lift some of the matters that were raised in the two Commission reports on ERIC, as well as at the biannual ERIC Forum meetings held between 2016 and 2018 and attended by the author.

Put forward via a directly applicable EU regulation, the legal-administrative form of ERIC is meant to be recognized in all of the EU member states. An advantage of the ERIC framework is an easier set-up of multinational RI, particularly when compared with lengthy and complex ratification processes required for a treaty establishing an intergovernmental scientific organization under international public law (OECD 2014: 12). Another advantage which the ERIC Regulation grants are certain benefits that an RI obtains together with the legal status of ERIC. These include VAT exemption, freedom to craft its own internal policies, including those on public procurement that allows to surpass the cumbersome EU procurement legislation. Most importantly, an RI attains a legal personality under the status of ERIC that implies liability and a possibility to enter into agreements with other societal actors (e.g. partners, contractors, etc.).

When obtaining a legal personality, an ERIC organization also obtains a mix of three jurisdictions. A facility with ERIC status is governed by the ERIC Regulation, the law of the state where the statutory seat of the RI is located, as well as its statutes and implementing rules. Such an arrangement creates a hybrid legal-administrative form for pan-European RIs (Denis et al 2015: 275). The hybridity is particularly evident due to ERIC being neither an EU agency nor an IO in the traditional sense (Reichel et al 2014: 1056). A certain hybridity of ERIC is also evident in view of its EU/IO characteristics when it comes to public procurement (Graber-Soudry 2019). Since most of the established ERICs are distributed RIs, the presence of nodes in various countries may pose certain legal challenges when it comes to their functioning as a multitude of existing organizations, united under the ERIC "umbrella".

The crossbreed nature of an entity established under the ERIC form may experience challenges in performing its organizational daily functions. The questions of liability and recognition by such actors as corporate financial institutions or contractors constitute additional challenges for ERICs. Some national registries were reported to struggle with determining under what category this type of legal entity ERIC falls and how it should be registered by the national tax authorities (European Commission 2014b, 2018). Hence, it becomes apparent that even though the ERIC framework was designed with a "one size fits all" arrangement in mind, the diverse missions, applications and uses of the individual ERIC organizations require different rationales and solutions when drafting their statutes and internal policies (Moskovko et al 2019).

While ERIC may be seen as an achievement for the European scientific collaboration, the legal form does not yet appear feasible for accommodating non-European countries as equal partners and shareholders in a multinational RI. It is required that an ERIC must be composed of at least one EU member state and two other countries that can be either member states or associated countries (European Council 2009). Initially this requirement applied only to the member states and having at least three of them to set up and operate an ERIC. In 2013 the ERIC Regulation was amended to better reflect the contribution of associated countries to the ERA and give them a better opportunity of participating in European RIs (European Council 2009, Amendment, Preamble and Article 1).

In its first report on ERIC, the European Commission stated that "associated countries or third countries to which the ERIC Regulation is not directly applicable [...] need to submit a binding declaration recognising the legal personality and the privileges of an ERIC for possibly hosting (in the case of associated countries) or becoming a member of a specific ERIC" (European Commission 2014b: 5). So far only Israel and Serbia have submitted those declarations (European Commission 2018: 7). Norway went through a rather quick parliamentary procedure lasting two months in order to recognize ERIC within its legislature (Stortinget 2015a, 2015b) and subsequently became the host of two ERICs.

Some of the ERICs established to date have also demonstrated a practice of including IOs among their members: EU-OPENSCREEN has the EMBL among its partners, while the Dutch Language Union is listed as a member of CLARIN. The Cherenkov Telescope Array that intends to apply for ERIC status is to have ESO among its members. Private parties such as companies cannot be members of an ERIC, however, ERICs may enter into agreements with "third parties". To date, CLARIN – an RI in the area of linguistics – has a formal agreement with a United States-based organization that contributes some of the American scientific resources to the European RI. The framework however is not suitable for American participation due to its concerns about sovereignty and the fact that ERIC members are to submit to the EU jurisdiction (Moskovko and Teich forthcoming).

It is important to note that the voting rights do not necessarily correspond to members' input as member states or associated countries, or a mixture of the two, and shall at all times jointly hold majority voting (European Council 2009, Article 9(3)). Thus, irrespective of third countries or intergovernmental organizations making financial contributions to an ERIC, the power to decide on the essential matters of the ERIC should still be in the hands of the EU member countries, or nations closely associated with the EU, in accordance with "the Community dimension of this Regulation" (European Council 2009, Preamble, Recital (14)). On a similar note, the statutory seat of the ERIC legal entity shall be in a member state or associated country (European Council 2009, Article 8(1)). Therefore, despite the good intentions of the enacted legal form for European science, some of the challenges remain for the "beyond Europe" dimension and are yet to be worked out by the EU and member states.

It also remains to be seen how the uncertainty that came with the 2016 United Kingdom (UK) referendum on exiting the EU (Brexit) may also affect the ERIC landscape when the EU treaties no longer apply to the UK, and the EC implementing decisions on the establishment of the individual ERICs that would be in effect only as long as the UK is a member state (European Commission 2018: 6 with reference to Article 50 TEU 2012). The UK is currently host of two ERICs and a participating member of nine (House of Commons 2018). Interestingly, in view of the Brexit referendum which was approaching at the time, the UK became an official member of the European Spallation Source in 2016 (UKRI 2016) and a host to the Integrated Structural Biology Infrastructure (INSTRUCT) in 2017. Hence, the EC decision establishing INSTRUCT stated that it would only apply until the UK ceased to be an EU member. It was also considered that in case the UK "ceases to be a Member State and without prejudice to the provisions of a possible withdrawal agreement headquarters of INSTRUCT ERIC would be related to the territory of another EU member state or an associated country" (European Commission 2017, Preamble 2, 3).

Despite the problems, Brexit may have also shed light on the matter of ERIC on the EU-wide arena. On November 25, 2018, the political declaration setting out the framework for the future relationship between the EU and the UK explicitly stated an intention for the parties to explore how the UK would participate in the EU science and innovation programs and specifically the ERIC (European Commission 2019f , Section A(12)). In view of that, parliamentary debates were held on the matter of ERIC and Brexit at both chambers of the UK Parliament that stressed that the nature of ERIC would not affect the sovereignty of the UK after Brexit and that the UK should continue its involvement in the ERICs (House of Commons 2018; House of Lords 2018). Moreover, the limited role of the EU in ERICs was pointed out on the basis of the countries (and not the EU) being the funders of ERICs and since penalties could not be imposed for non-compliance by the European Court of Justice. Instead, it was stated that each ERIC should be approached on a case-by-case basis "in line with its own rules – rules to which members of the consortium agree when they sign up" (House of Commons 2018).

The above once again demonstrates the "plasticity" of the ERIC form, able to suit both Euro-optimists and Euro-skeptics. The section above has also demonstrated the ability of ERIC to accommodate associated countries and IOs as members and to enter into contractual agreements with "third parties" (such as own RI of an American university that is now integrated into the shared resources of CLARIN). Even though the matters of taxation and public procurement are yet to be worked out for ERICs, the Brexit discussions on this European framework have demonstrated that ERIC has gained a certain degree of stability and momentum. It may also be posited that its perceived benefits are visible even to an EU member state that is set to leave the EU. Hypothetically, ERIC as a legal form may be amended even further, possibly also once the initial cohort of the ERICs matures and enters the consecutive phases of their respective life cycles, or provided new needs and challenges arise. Without doubt, being a legal act of the EU, the ERIC form may also get cancelled altogether. However, the latter scenario is less likely considering the momentum that the ERIC instrument has gained so far.

5. CONCLUDING REMARKS

This chapter has presented ERIC from three perspectives: how it came about as a response to an "elevated" attention to the RI in view of the ERA, what characteristics are shared by the 21 ERICs established to date and how the policy tool of ERIC has been applied during the last decade. It has been stressed that despite ERIC being a product of practical application of EU law – through the full use of Article 187 of the Treaty on the Functioning of the European Union in order be able to utilize its power to create ERIC as "any other structure" – the overall involvement of the EU into the matter of ERIC remains rather marginal. While the European Commission is a primary reviewer of the applications for ERIC status and an advisor on the implementation process, the list of the responsibilities of the EU around the ERIC ends there. The underlying reason is of course the limitations that remain in place, most of all the area of research and innovation occupying an area of shared responsibilities between the EU and the member states (TFEU 2012, Article 4). Moreover, the wording of the ERIC Regulation explicitly states that ERIC *should not be perceived as an EU body*, since the member states hold joint responsibility and liability for it.

Nonetheless, being a secondary EU legislation, ERIC is susceptible to change. This has already been demonstrated by the 2013 amendment to the ERIC Regulation that lifted up the importance of the associated member countries and allowed them to host ERICs. The "beyond Europe" dimension of ERIC may require extra time to get figured out – the involvement of the non-European countries is currently subject to shortcomings in the voting rights (European Council 2009, Article 9(2)), as well as the EU jurisdiction (European Council 2009, Article 15). The question of funding also inexplicitly lures on the background – would the EU or the participating member states be responsible and hence liable for a global endeavor that runs under the ERIC framework?

There are currently 21 research organizations that function as ERICs. Out of these, only one is a "Big Science" facility, while the rest may be classified as distributed resources under the ERIC umbrella. Nineteen of those RIs are included in the list of strategic pan-European projects by ESFRI. While both center and periphery of the EU are well represented among the EU participants, the statutory seats of the ERICs are predominantly located in the core. An observation may be made that most of the distributed ERICs were set up following the "preparatory phase funding" initiatives available under the FP to projects prioritized by ESFRI. While a certain momentum has been gained, it is not completely clear what the future of ERIC facilities looks like. How sustainable is the ESFRI-infused model for prioritizing initiatives, making "preparatory funding available" and then producing ERIC structures? How can a balanced distribution of ERICs be sustained in the EU center and periphery? These are important questions for future research.

A certain degree of institutionalization of ERIC has been witnessed through the momentum that the legal framework has gained over the years: the 21 ERICs established to date, the 2013 amendment of the legal form that demonstrated the "plasticity" of this policy instrument and stimulated the engagement of the associated countries and the formation of the ERIC Forum where best practices are shared and collaborative work undertaken to solve common challenges. Moreover, despite the turbulence caused by Brexit, the ERIC landscape persists. Brexit may as well have raised awareness and drawn attention to ERIC, at the same time pointing out the resilience of this policy instrument. Being a legislative act of the EU, the ERIC regulation may be susceptible to change and may also be altered or called off altogether. ERIC, however, is most likely here to stay.

Overall, ERIC being an EU creation may be seen as an outcome of the intensified role of the EU's research and innovation policy, streamlined with other ambitious policy goals of the EU. With the €2.5 billion proposed just for RIs in the next FP budget, there is a signal for even more opportunities for constructing novel RIs. ERIC, considering the gained momentum, might continue to be a preferred way to establish those. The question remains, however, whether the EU will get more involved in the funding of operations of the currently established ERICs (particularly those in the EU "periphery"), and how new challenges could be solved when the "older cohort" of ERICs move on to the next stages of their respective life cycles. Would the ERIC in its present shape and form be able to accommodate the challenges that the evolving scientific disciplines, methods and technologies are calling for?

A certain stability has been reached and the initial shortcomings have been solved for the ERIC framework between the member states and the EU (European Commission 2014b, 2018). However, the question of VAT exemption for in-kind contributions are yet to be settled for the current and upcoming ERICs. It may be stated that ERICs will continue to occupy an undefined legal space at the juncture of EU law, national jurisdiction and ERIC-specific statutes and policies. The last ten years have demonstrated that ERIC has matured and proved a usable policy instrument. The next decade may bring up an even more increased institutionalization and diffusion of this legal form, possibly becoming a stepping stone for creating "global distributed RIs" (OECD 2014).

ACKNOWLEDGMENTS

Maria Moskovko would like to thank the individuals who kindly shared their knowledge on and experience with the ERIC framework via personal interviews and/or conversations during the ERIC Forum meetings that she attended between 2016 and 2018.

REFERENCES

Barton O (2014) Analysis of the principle of subsidiarity in European Union law. *North East Law Review* **2**: 83–8.

BMBF (2019) Email correspondence with a representative of the German Federal Ministry of Education and Research.

Buonanno L and N Nugent (2013) *Policies and policy processes of the European Union.* Palgrave Macmillan.

Chou M-H (2014) The evolution of the European Research Area as an idea in European integration. In M-H Chou and Å Gornitzka (eds) *Building the knowledge economy in Europe: New constellations in European research and higher education governance.* Edward Elgar Publishing, pp 27–50.

Delanghe H, U Muldur and L Soete (eds) (2009) *European science and technology policy: Towards integration for fragmentation?* Edward Elgar Publishing.

Denis J, E Ferlie and N van Gestel (2015) Understanding hybridity in public organizations. *Public Administration* **93** (2): 273–89.

ECRI (2006) *Third European Conference on Research Infrastructures Nottingham, UK – 6 and 7 December 2005. Main conclusions.* Office for Official Publications of the European Communities (last accessed January 10, 2017).

ECRI (2007) *The Fourth European Conference on Research Infrastructures. Challenge 4a: Legal aspects.* Hamburg, June 5–6. Available at: www.ecri2007.de/programme/challenge_4/challenge_4a/index_eng.html (last accessed May 1, 2017).

ERIC Forum (2017) *First ERIC Forum meeting.* University of Graz, November 16–17. Attended by the author.

ERIC Forum (2018) *Third ERIC Forum meeting.* Escuela de Organización Industrial, Seville, November 27–28. Attended by the author.

ERIC Forum (2019) *ERIC Forum members' written contribution to ERAC ad-hoc working group on the Future of the ERA*. Available at: www.eric-forum.eu/2019/09/16/eric-forum-members-written-contribution-to-erac-ad-hoc-working-group-on-the-future-of-the-era/ (last accessed December 6, 2019).

ESFRI (2006) *European Strategy Forum on Research Infrastructures roadmap for Research Infrastructures 2006*. Available at: www.esfri.eu/sites/default/files/esfri_roadmap_2006_en.pdf (last accessed July 26, 2019).

ESFRI (2008) *European Strategy Forum on Research Infrastructures roadmap for Research Infrastructures update 2008*. Available at: www.esfri.eu/sites/default/files/esfri_roadmap_update_2008.pdf (last accessed July 27, 2019).

ESFRI (2016) *Strategy report on Research Infrastructures: Roadmap 2016*. Available at: www.esfri.eu/esfri_roadmap2016/roadmap-2016.php (last accessed November 15, 2019).

ESFRI (2018) *European Strategy Forum on Research Infrastructures roadmap 2018 – Strategy report on Research Infrastructures*. Available at: http://roadmap2018.esfri.eu/media/1060/esfri-roadmap-2018.pdf (last accessed November 15, 2019).

European Commission (2008a) *Commission staff working document COM (2008) 467 final. Accompanying document to the proposal for a council regulation on the community legal framework for a European Research Infrastructure (ERI). Impact assessment*. Available at: https://eur-lex.europa.eu/LexUriServ/LexUriServ.do?uri=SEC:2008:2278:FIN:EN:PDF (last accessed August 22, 2018).

European Commission (2008b) *Proposal for a council regulation on the community legal framework for a European Research Infrastructure (ERI) 2008/0148*. European Commission. Available at: http://register.consilium.europa.eu/doc/srv?l=ENandf=ST%2012259%202008%20INIT (last accessed February 1, 2019).

European Commission (2008c) *Developing world-class Research Infrastructures for the European Research area (ERA): Report of the ERA expert group*. Office for Official Publications of the European Communities. Available at: http://ec.europa.eu/research/era/pdf/dgs138-era-expert-group-final-low-080212_en.pdf (last accessed February 1, 2019).

European Commission (2011) *Commission decision 2011/166/EU of 17 March 2011 setting up the SHARE-ERIC. Amended 31/05/2014*. Available at: https://eur-lex.europa.eu/legal-content/EN/TXT/?uri=CELEX%3A32011D0166 (last accessed November 15, 2019).

European Commission (2014a) *Commission implementing decision 2014/392/EU of 24 June 2014 on setting-up the Central European Research Infrastructure Consortium (CERIC-ERIC)*. Available at: https://eur-lex.europa.eu/legal-content/EN/TXT/?uri=CELEX%3A32014D0392 (last accessed November 15, 2019).

European Commission (2014b) *Report from the Commission to the European Parliament and the Council on the application of council regulation (EC) No 723/2009 of 25 June 2009 on the community legal framework for a European Research Infrastructure consortium (ERIC)*. Available at: https://ec.europa.eu/research/infrastructures/pdf/eric_report-2014.pdf#view=fitandpagemode=none (last accessed February 1, 2019).

European Commission (2017) *Commission implementing decision (EU) 2017/1213 of 4 July 2017 on setting up the Integrated Structural Biology — European Research Infrastructure Consortium (Instruct-ERIC)* (notified under document C (2017) 4507). Available at: https://eur-lex.europa.eu/legal-content/EN/TXT/?uri=CELEX%3A32017D1213 (last accessed November 15, 2019).

European Commission (2018) *Report from the Commission to the European Parliament and the Council: Second report on the application of council regulation (EC) No 723/2009 of 25 June 2009 on the Community legal framework for a European Research Infrastructure Consortium (ERIC)*. Available at: https://eur-lex.europa.eu/legal-content/EN/TXT/PDF/?uri=CELEX:52018DC0523andfrom=EN (last accessed February 1, 2019).

European Commission (2019a) *Realising and managing international research infrastructures – RAMIRI 2008–2010*. Cordis EU research results. Grant agreement ID: 226446. Last update: July 16, 2019. Available at: https://cordis.europa.eu/project/id/226446 (last accessed December 16, 2019).

European Commission (2019b) *Realising and managing international Research Infrastructures 2 – RAMIRI 2 2010–2013*. Cordis EU research results. Grant agreement ID: 262567. Available at: https://cordis.europa.eu/project/id/262567 (last accessed December 16, 2019).

European Commission (2019c) *Research Infrastructures Consortium for Horizon 2020 – RICH 2014–2019*. Cordis EU research results. Grant agreement ID: 646713. Available at: https://cordis.europa.eu/project/id/646713 (last accessed December 16, 2019).

European Commission (2019d) *Research Infrastructures training programme*. Cordis EU research results – RItrain. Grant agreement ID: 654156. Available at: https://cordis.europa.eu/project/id/654156 (last accessed December 16, 2019).

European Commission (2019e) *ERIC Forum implementation project – ERIC Forum*. Grant agreement ID: 823798. Available at: https://cordis.europa.eu/project/id/823798 (last accessed December 16, 2019).

European Commission (2019f) *Revised text of the political declaration setting out the framework for the future relationship between the European Union and the United Kingdom as agreed at negotiators' level on 17 October 2019, to replace the one published in OJ C 66I of 19.2.2019*. Available at: https://ec.europa.eu/commission/publications/revised-political-declaration_en (last accessed November 15, 2019).

European Commission (2019g) *Horizon EUROPE: The next EU Research and Innovation Investment Programme (2021–2027)*. May. Version 25. Available at: https://ec.europa.eu/info/horizon-europe-next-research-and-innovation-framework-programme_en (last accessed August 26, 2019).

European Council (1985) *Council regulation (EEC) No 2137/85 of 25 July 1985 on the European Economic Interest Grouping (EEIG)*. Available at: https://eur-lex.europa.eu/legal-content/EN/ALL/?uri=celex%3A31985R2137 (last accessed November 15, 2019).

European Council (1987) *Council decision of 28 September 1987 concerning the framework programme for community activities in the field of research and technological development (1987 to 1991) (87/ 516/Euratom, EEC)*. Available at: https://eur-lex.europa.eu/legal-content/EN/TXT/?uri=CELEX%3A31987D0516 (last accessed July 25, 2019).

European Council (1990) *90/221/Euratom, EEC: Council decision of 23 April 1990 concerning the framework programme of community activities in the field of research and technological development (1990 to 1994)*. OJ L 117, May 8, pp 28–43. Available at: https://eur-lex.europa.eu/legal-content/EN/TXT/?uri=CELEX%3A31990D0221 (last accessed December 16, 2019).

European Council (2001) *Council regulation (EU) No 2157/2001 of 8 October 2001 on the statute for a European company (SE)*. Available at: https://eur-lex.europa.eu/

LexUriServ/LexUriServ.do?uri=CELEX:32001R2157:EN:HTML (last accessed January 10, 2020).
European Council (2002) *Council regulation (EC) No 876/2002 of 21 May 2002 setting up the Galileo joint undertaking*. Available at: https://eur-lex.europa.eu/legal-content/EN/TXT/?uri=CELEX:32002R0876 (last accessed November 15, 2019).
European Council (2009) *Council regulation (EC) No 723/2009 of 25 June 2009 on the community legal framework for a European Research Infrastructure Consortium (ERIC) [2009] OJ L 206/1, as later amended by council regulation (EC) No 1261/2013 of 2 December 2013 [2013] OJ L 326/1*. Available at: https://eur-lex.europa.eu/legal-content/EN/TXT/PDF/?uri=OJ:L:2013:326:FULLandfrom=ES (last accessed November 15, 2019).
European Council (2013) *Council regulation (EU) No 1261/2013 of 2 December 2013 amending Regulation (EC) No 723/2009 concerning the community legal framework for a European Research Infrastructures Consortium (ERIC)*. Available at: https://eur-lex.europa.eu/legal-content/EN/TXT/?uri=CELEX%3A32013R1261 (last accessed November 15, 2019).
European Parliament (1994) *Decision No 1110/94/EC of the European Parliament and of the Council of 26 April 1994 concerning the fourth framework programme of the European Community activities in the field of research and technological development and demonstration. OJ L 126*, May 18, pp 1–33. Available at: https://eur-lex.europa.eu/legal-content/PL/TXT/?uri=CELEX:31994D1110 (last accessed December 16, 2019).
European Parliament (1999) *Decision No 182/1999/EC of the European Parliament and of the Council of 22 December 1998 concerning the fifth framework programme of the European Community for research, technological development and demonstration activities (1998 to 2002)*, OJ L 26, February 1, pp 1–33. Available at: https://eur-lex.europa.eu/legal-content/EN/ALL/?uri=CELEX:31999D0182 (last accessed December 16, 2019).
European Parliament (2000) *Lisbon European Council 23 and 24 March 2000. Presidency conclusions*. Available at: www.europarl.europa.eu/summits/lis1_en.htm (last accessed July 29, 2019).
European Parliament (2006) *Regulation (EC) No 1082/2006 of the European Parliament and of the Council of 5 July 2006 on a European grouping of territorial cooperation (EGTC)*. Available at: https://eur-lex.europa.eu/legal-content/EN/ALL/?uri=CELEX:32006R1082 (last accessed November 15, 2019).
European Parliament (2013) *Regulation (EU) No 1291/2013 of the European Parliament and of the Council of 11 December 2013 establishing Horizon 2020 – the Framework Programme for Research and Innovation (2014–2020) and repealing decision no 1982/2006/EC text with EEA relevance*. Available at: https://eur-lex.europa.eu/legal-content/EN/TXT/?uri=CELEX%3A32013R1291 (last accessed November 15, 2019).
European Parliament (2016) *The principle of subsidiarity, fact sheets on the European Union*. Available at: www.europarl.europa.eu/atyourservice/en/displayFtu.html?ftuId=FTU_1.2.2.html (last accessed October 1, 2017).
George A and A Bennett (2005) *Case studies and theory development in the social sciences*. MIT Press.
Graber-Soudry O (2019) Regulating procurement by European Research Infrastructure Consortia (ERICs) and the exemption for international organisations. *Studies of the Oxford Institute of European and Comparative Law* **26**: 249–64.

Guston D H (2001) Boundary organizations in environmental policy and science: An introduction. *Science Technology and Human Values* **26** (4): 399–408.
Hallonsten O (2014) The politics of European collaboration in Big Science. In M Mayer, M Carpes and R Knoblich (eds) *The global politics of science and technology*, Vol. 2. Springer-Verlag, pp 31–46.
Hilling A, O Hallonsten and E Engholm (2017) Tax policy issues in connection with the European Spallation Source Project and other European Research Infrastructure Consortiums. *Skattepolitisk Oversigt* **8**: 1–9.
House of Commons (2018) European Statutory Instruments Committee. First report of session 2017–19. Documents considered by the committee on 5 September 2018. HC1532. September 7. Available at: https://publications.parliament.uk/pa/cm201719/cmselect/cmesic/1532/153202.htm (last accessed December 1, 2018).
House of Lords (2018) Appendix 3: Draft European Research Infrastructure Consortium (Amendment) (EU exit) regulations 2018. November 7. Available at: https://publications.parliament.uk/pa/ld201719/ldselect/ldseclega/221/22109.htm (last accessed December 1, 2018).
ICRI (2018) Personal communication during International Conference on Research Infrastructure. September 12–14. Federal Ministry of Education, Science and Research. Vienna.
Krige J (2003) The politics of European scientific collaboration. In J Krige and D Pestre (eds) *Companion to science in the twentieth century*. Routledge.
Moskovko M and A Teich (Forthcoming) *Science diplomacy in action: European collaboration and US participation in Research Infrastructures*.
Moskovko M, A Astvaldsson and O Hallonsten (2019) Who is ERIC? The politics and jurisprudence of a new governance tool for collaborative European Research Infrastructures. *Journal of Contemporary European Research* **15** (3): 249–68.
MusicNOW (2018) MusicNOW to drive European Commission application for an ERIC MusicNOW website. Available at www.kenup.eu/MusicNow (last accessed December 1, 2018).
Öberg J (2016) Subsidiarity as a limit to the exercise of EU competences. *Yearbook of European Law*: 1–30.
OECD (2014) *International distributed Research Infrastructures*. Report. OECD Global Science Forum. Available at: www.oecd.org/sti/sci-tech/international-distributed-research-infrastructures.pdf (last accessed November 13, 2016).
Papon P (2004) European scientific cooperation and Research Infrastructures: Past tendencies and future prospects. *Minerva* **42**: 61–76.
Papon P (2009) Intergovernmental cooperation in the making of European research. In H Delanghe, U Muldur and L Soete (eds) *European science and technology policy: Towards integration for fragmentation?* Edward Elgar Publishing, pp 24–43.
Reichel J, A-S Lind and M Hansson (2014) ERIC: A new governance tool for biobanking. *European Journal of Human Genetics* **22**: 1055–7.
Ryan, L (2015) Governance of EU research policy: Charting forms of scientific democracy in the European Research Area. *Science and Public Policy* **42**: 300–14.
Ryan, L (2019) Balancing rights in the European Research Area: The case of ERICs (European Research Infrastructure Consortium). *European Intellectual Property Review* **4** (41): 218–27.
Schiermeier, Q (2000) Europe urged to set up advisory body on research infrastructure. *Nature* 407: 433–34.

SEA (1986) Single European Act. OJ L 169, 29.6.1987, 1–28. Available at: https://eur-lex.europa.eu/legal-content/EN/TXT/?uri=CELEX:11986U/TXT (last accessed October 10, 2019).

SHARE (2016) Survey of Health, Ageing and Retirement in Europe (SHARE). Frequently asked questions, SHARE website. Available at: www.share-project.org/?id=762#1.1 (last accessed January 10, 2017).

Star S and J Griesemer (1989) Institutional ecology, "translations", and boundary objects: Amateurs and professionals in Berkeley's Museum of Vertebrate Zoology 1907–39. *Social Studies of Science* **19**: 387–420.

Stortinget (2015a) Samtykke til godkjennelse av EØS-komiteens beslutning om innlemmelse i EØS-avtalen av forordning om Fellesskapets rettslige ramme for et konsortium for en europeisk forskningsinfrastruktur (ERIC-konsortium). Available at: www.stortinget.no/no/Saker-og-publikasjoner/Saker/Sak/?p=62369 (last accessed December 5, 2018).

Stortinget (2015b) Lov om konsortium for europeisk forskingsinfrastruktur (ERIC-lova). Available at: www.stortinget.no/no/Saker-og-publikasjoner/Saker/Sak/?p=63390 (last accessed December 5, 2018).

TFEU (2012) Consolidated version of the Treaty on the Functioning of the European Union. OJ C 326, October 26, pp 47–390. Available at: https://eur-lex.europa.eu/legal-content/EN/TXT/?uri=celex%3A12012E%2FTXT (last accessed December 16, 2019).

Treaty of Lisbon (2007) Treaty of Lisbon amending the Treaty on European Union and the Treaty establishing the European Community, signed at Lisbon, 13 December 2007. OJ C 306, December 17, pp 1–271. Available at: https://eur-lex.europa.eu/legal-content/EN/TXT/?uri=CELEX%3A12007L%2FTXT (last accessed October 10, 2019).

Treaty on European Union (2012) Consolidated version of the Treaty on European Union. OJ C 326, October 26, pp 13–390. Available at: https://eur-lex.europa.eu/legal-content/EN/TXT/?uri=celex%3A12012M%2FTXT (last accessed December 16, 2019).

UKRI (2016) UK confirmed as founding member of world's largest microscope, June 28. Available at: https://stfc.ukri.org/news/uk-confirmed-as-founding-member-of-world-s-largest-microscope/ (last accessed November 15, 2019).

Yu H, J B Wested and T Minssen (2017) Innovation and intellectual property policies in European Research Infrastructure Consortia. Part I: The case of the European Spallation Source ERIC. *Journal of Intellectual Property Law and Practice* **12** (5): 384–97.

Zapletal, J (2010) The European Grouping of Territorial Cooperation (EGTC): A new tool facilitating cross-border cooperation and governance. *Quaestiones Geographicae* **29** (2): 15–26.

7. Research Infrastructure funding as a tool for science governance in the humanities: A country case study of the Netherlands

Thomas Franssen

1. INTRODUCTION

While the bulk of Research Infrastructures (RIs) in Europe are built for the natural sciences, the humanities have not been left behind. Both on the European level, for example, through the inclusion of projects in the European Strategy Forum on Research Infrastructures (ESFRI) roadmaps and within different national contexts, there has been considerable enthusiasm for the development of RIs in the humanities. The Netherlands is a good example. After a relatively slow start in the early 2000s, when the Netherlands was seen as having to catch up on RI development, the past 15 years has seen an increase in funding across all scientific domains. In the humanities, ten projects with a total budget of €50 million have been funded since 2003. This chapter aims to contextualize and critically analyze this funding boom as an example of the general upswing that RI funding appears to have had in many European countries (see Bolliger and Griffiths, ch 5 in this volume), and thus studying RI funding as an instrument for science governance in the humanities.

RI projects included on the ESFRI roadmap are very heterogeneous, even to the extent that Hallonsten (2020: 11) argues that the roadmap should not be understood to represent a particular mode of science, but is rather a reflection of a political process. However, within the humanities, RI projects on the ESFRI roadmap are surprisingly homogeneous: they are all multipurpose, multisite repositories (in the terminology developed by Hallonsten 2020), and all have a strong digital component. I, therefore, characterize them in this chapter as digital RIs. This characterization also makes clear that infrastructure development in the humanities cannot be understood separately from the rise of digital humanities as a research area. Indeed, the institutionalization of digital humanities and the success of digital RI proposals go hand in hand.

This chapter highlights the relation between RI funding and science policy goals. While RIs in the humanities are not "Big Science" in the narrow sense (see Cramer et al, ch 1 in this volume) as defined through "big machines" and "big organizations", it does involve "bigger" politics. The investments in RIs in the humanities are tied to big expectations in science policy. The humanities have – in the Dutch science policy context – struggled. The epistemic fragmentation of the humanities (Whitley 1984) is seen as problematic, and their societal value, evaluated in light of the perceived demands of the labor market, has been questioned ever since the 1980s. In contrast, the development of digital RIs in the humanities is seen to be a solution to this fragmentation as it is expected to increase collaboration in the domain as well as change the organization and substance of research done in this area.

The chapter is structured in the following way: First, I outline the body of literature that informs my understanding of the relation between science governance and research practices, and the tools available to governance actors, such as funding arrangements, which mediate science policy aims. Second, I contextualize the rise of funding for research infrastructure in the humanities by telling three histories: (1) the main aims of Dutch science policy and the position of the humanities within it, (2) the rise of digital humanities in the Netherlands and (3) the rise of funding arrangements for RIs. Third, I describe the funding boom for RIs in the humanities, specifically drawing attention to various early infrastructural projects and, most importantly, the Common Lab Research Infrastructure for the Arts and Humanities (CLARIAH). This RI project is the biggest collaboration in the field, drawing together the communities around the Common Language Resources and Technology Infrastructure (CLARIN) and Digital Research Infrastructure for the Arts and Humanities (DARIAH), and is seen in science policy as the defining digital RI for the humanities. Finally, I draw upon an analysis of science policy documents published or commissioned by the Ministry of Education, Culture and Science (Ministerie voor Onderwijs, Cultuur en Wetenschap, abbreviated OCW), the Royal Netherlands Academy of Arts and Science (Koninklijke Nederlandse Akademie van Wetenschappen, abbreviated KNAW), the Dutch Science Funding Organization (Nederlands Organisatie voor Wetenschappelijk Onderzoek, abbreviated NWO) and the Advisory Council for Science, Technology and Innovation (Adviesraad voor Wetenschap, Technologie en Innovatie, abbreviated AWTI) in the last 15 years to show the science policy expectations around CLARIAH. To confirm and contextualize the analysis, I interviewed two policymakers at the Ministry of Education, Culture and Science. In the conclusion I reflect on the feasibility of these expectations.

2. RESEARCH INFRASTRUCTURE FUNDING AS A TOOL FOR SCIENCE GOVERNANCE

A crucial question in the sociology of science, science policy studies and science and technology studies is the extent to which science governance actors – such as national governments, funding bodies and research organizations – can and should steer the content of research. Answering this question has been hampered by a disciplinary split between those that mainly study the governance of science and those that study research practices (Gläser and Laudel 2016). An emerging body of literature that aims to (re)connect these domains attempts to theorize the ways in which science governance actors try to influence the content of science in relation to the ways in which the content of science is determined by researchers themselves.

We know from past research that the strategic capabilities of science governance actors are limited (Whitley 2008; Gläser 2019). It is extremely difficult to steer research content because of the high level of uncertainty regarding what research problem should be prioritized and then deciding how it should be solved. Science governance actors do not usually have the competences to dictate a particular course of action. Moreover, many decisions are made during the research process; this often impedes the possibility of initially determining which research problem will be – and how it will be – studied. Thus, rather than direct steering, science governance actors try to influence how researchers do research.

In the sociology of science and science and technology studies, research has studied the ways research problems are defined in the context of the possibilities researchers have, so-called "doable" problems (Fujimura 1987), how research lines are developed in the context of the career system (Laudel 2017; Laudel and Bielick 2018) and how portfolios of research topics come to align with particular valuation regimes dominant in the science system (Rushforth et al 2019). A number of studies have also analyzed how particular forms of funding are used by researchers (Laudel and Gläser 2014; Franssen et al 2018; Whitley et al 2018).

In 2019, Gläser conceptualized four ways in which research organizations can try to influence research content: (1) by coercion through commanding or prescriptive rules; (2) by equipping through direct allocation or institutionalized allocation procedures; (3) by inducing through reward or institutionalized reward procedures; and (4) by suggesting reinterpretation of the situation through the transmission of specific information or through institutionalized belief systems.

Gläser (2019) argues that equipping through allocation of funding is often combined by what he calls the "reinterpretation of the situation". For instance, a university department might have a particular idea of what the "next big thing is" (e.g. artificial intelligence) and, accordingly, allocate funding to strengthen research on this particular topic in the organization (see also Laudel and Weyer 2014). Similarly, a government or funding body's idea of "good science" might change – for instance, in the support of open science or by signing the San Francisco Declaration on Research Assessment (DORA) – and can consequently adapt its allocation procedures to such an "institutionalized belief system" or epistemic regime (Elzinga 1993).

Drawing upon this literature, I argue that we should understand science policy discourses as the preferred ways of understanding the organization, goal and societal purpose of scientific research. A science policy discourse thus consists of particular normative frames regarding what "good science" is.

One of the ways in which such normative frames are brought into the vicinity of researchers is through funding arrangements. Funding arrangements, through which research is funded, have particular epistemic dimensions in terms of size, project length and evaluation criteria. It is, therefore, necessary to understand these as epistemic devices that reconfigure research practices in particular ways (Law and Ruppert 2013). Each funding arrangement has particular affordances and constraints in terms of what can or cannot – or better, should and should not – be done with it (Franssen et al 2018). Funding arrangements are thus "charged" with particular normative frames.

RI funding is no exception. As I will show, funding arrangements for RIs and instrumentation are charged with normative frames that are aligned with the main aims articulated in the Dutch science policy discourse. The success of the humanities in these funding arrangements thus reflects the strengthening of a particular way of organizing humanities research and of particular research traditions within the humanities.

3. SETTING THE STAGE: THREE HISTORIES

To understand the significance of recent investments in digital RIs in the Dutch humanities, I propose that we need to take into account three significant histories. The first is the ways in which the humanities have been qualified in Dutch science policy since the late 1970s. The second is the development of funding arrangements for RIs in the 1990s and 2000s. The third is the emergence and institutionalization of digital humanities as a research area since the 1980s.

3.1 The Fragmentation of the Humanities as a Science Policy Problem

An important recurring element of Dutch science policy since the late 1970s is the ambition to increase collaboration between researchers and coordination of research agendas through priority setting (Laudel and Weyer 2014; van der Meulen 2007; Blume and Spaapen 1988; Benneworth et al 2016). The increase of collaboration and coordination is meant to create critical mass around a limited number of focus points. This is deemed necessary because the science system is too small to produce good research in every research area; focusing on a limited number of research areas is assumed to lead to higher-quality research.

Departing from a more observational position in the 1960s and early 1970s, the Dutch government became increasingly involved in the coordination of research activities of scholars. Themed funding programs were introduced and disciplinary communities as well as universities were asked to plan and coordinate their research efforts (de Boer et al 2007). By the mid-1980s, accountability became integral as science governance actors looked for a more explicit relation between input and output, which introduced research evaluations into the Dutch university system (de Haan 1994). Moreover, the sector was expected to purposefully articulate the societal value of its research and teaching particularly in relation to the labor market. This curricula change in the humanities also influenced other areas; for instance, Benneworth, Gulbrandsen and Hazelkorn describe the transformation of the Spanish language and literature degree at Leiden University into a broader Latin American language and culture degree, as "reflecting labour market demands by also equipping students to deal with a foreign society as well as simply learning the language" (Benneworth et al 2016: 125).

Various policy programs, such as "Task allocation and concentration" (1982) and "Selective shrinkage and growth" (1986), championed the idea that critical mass around a limited number of research areas or in a limited number of departments was the best possibility for the Netherlands to reach a high-quality level in research. Under the influence of this ambitious policy, we saw the merger or disappearance of small subject fields and small departments in social science and humanities disciplines for which alternatives became available at other Dutch universities (Blume and Spaapen 1988; de Haan 1994: 74).

In light of fostering coordination between various researchers' research agendas, and to engage more directly in research with clear societal relevance and demand, the epistemic fragmentation that characterizes many of the social sciences and humanities (as well as its academic curriculum) was a barrier. Research efforts should become better coordinated and more explicitly tied to the labour market; the result was problematic.

The subsequent friction between what science policy demanded and what most humanities disciplines traditionally offered was evident in various reports written about the state of the humanities between 1990 and 2010. The early 1990s saw the publication of two reports on the humanities by the advisory council (RAWB 1990a, 1990b) and two national committees were installed to guide the humanities into the future (Benneworth et al 2016). These two committees, namely Staal (1991) and Vonhoff (1995), focused on the "small humanities" (the languages) and, specifically, the societal value of humanities research and teaching that was seemingly out of touch with the demands of the labor market. They aimed to articulate this societal value and proposed financial solutions in order to guarantee the future sustainability of the small humanities. Of course, the committees were critical of the policies of the 1980s that were attempting to streamline and rationalize humanities research and teaching. Priority setting in research through the concentration of funding, however, continued through repeated calls for greater collaboration to develop critical mass around certain independently chosen research areas.

In the 2010 report, "Focus and mass in scientific research: The Dutch research portfolio in international perspective", Van de Besselaar and Horlings (2010) reflected upon science policy in the 2000s and its continued emphasis on creating focus and critical mass in the Dutch science system. They did not include the social sciences and humanities in their report, due to a lack of data, but it is evident that the focus of Dutch science policy had not changed. Moreover, the increasing focus on (economic) valorization and the introduction of policies to aid in public-private partnerships in the 2000s increased the need to articulate the societal value of the humanities even further and in an increasingly narrow direction (Benneworth et al 2016). Two committees, Gerritsen (2002) and Cohen (2008), again wrote reports about the humanities and its values and perspectives in light of these policy demands.

While the Gerritsen report did not become influential because of political upheaval, the Cohen report was followed by the establishment of a committee that implemented some of the recommendations made in this report, including greater regulation of the language studies portfolios of the Dutch universities. This committee received a budget of about €15 million a year. Most of this funding was distributed across humanities faculties and was used to solve local budgetary problems and make small investments. Benneworth et al (2016) summarize and argue that much has changed in terms of societal engagement in the research practices of humanities scholars. However, they also show that in the science policy discourse, to paraphrase, the humanities at large has been viewed as being in crisis, albeit without collapse, for the past 30 years:

> The issues at the end of the period [1982–2012] appear to be precisely the same as those at the start, a fragmented humanities field, too introverted and unable to steer itself to produce value for Dutch society or capable of being held accountable to government for its excellence. At least that is the impression that one would get if one's view of the public view of humanities was formed entirely by what elites said about humanities rather than to what humanities researchers were actually doing. (Benneworth et al 2016: 135)

3.2 The Emergence of (Competitive) Funding Arrangements for Research Infrastructure

Most of the funding for instrumentation and other technical resources in the 1980s was encapsulated in the block funding stream received by universities (Versleijen 2007). This changed as competitive funding arrangements became more popular across the science system. The change included funding arrangements for instrumentation and, later on, also for RI projects.

By the 1990s, the Dutch funding organization NWO developed a funding arrangement called NWO Investment Grant Large. At that time the funding arrangement was specifically meant to fund narrowly defined instrumentation. Project proposals aimed at funding had to have a budget above €900 000 (the total budget at that time was approximately €13.5 million per round). The scope needed to exceed individual institutions. The instrumentation needed to be accessible country-wide and its use had to foster national research priorities. The funding arrangement thus assumed (inter)disciplinary collaboration and coordination of research agendas across departments at different universities and among (sub)disciplines. These evaluation criteria matched the science policy's goal to increase collaboration and coordination as detailed above.

NWO Investment Grant Large was initially aimed at instrumentation that did not include, for instance, digitization of collections. It was only in 1999 that the funding arrangement was broadened and came to recognize the humanities and social sciences data collection efforts as infrastructural technology, writing: "[N]ext to acquisition [of instrumentation] the start and development of data-collection, as well as accompanying software and bibliographies is included if it makes an evident and nationally accessible contribution to the infrastructure" (NWO 1999). This was an important change in the funding arrangement and meant that instrumentation funding became available across a large number of scientific domains. Large instrumentation became, more generally, increasingly in focus within Dutch science policy at the turn of the millennium. A RAND Europe report for two ministries was published in 2001. The report's research subject, "scientific instrumentation", aimed to answer whether additional large instrumentation investments in the Dutch science system were needed. It was concluded that there was a projected need in the science system for investments in instrumentation (which included larger infrastructural technology) of an estimated amount of 3000 million guilders (or €1500 million).

With the establishment of the ESFRI in 2002, investment in RIs quickly became important in Dutch science policy. The preparations for the ESFRI roadmap, started in 2004 (Bollinger and Griffiths, ch 5 in this volume), gave a reason to think about the Dutch participation in these roadmap projects. A working group on "large-scale research infrastructures" was established and released an influential report in 2005. They argued, echoing the RAND Europe report, that the Netherlands was lagging behind in its involvement in RI development and that significant funds had to be made available in the coming years to bring the Netherlands into a leading position in European RI developments. While the resulting investments were smaller than the work group requested, nowhere near the €1500 million that RAND Europe argued was needed, there had been ongoing investments in RI projects of various sizes (beyond the NWO Investment Grant Large) since 2005.

An immediate investment was made to fund five projects in a so-called NWO-BIG round that included a €12 million project for the large-scale digitization of newspapers by the Dutch Royal Library (Versleijen 2007: 63). In addition to this, the development of a Dutch roadmap for RIs was proposed. The roadmap criteria were largely the same as ESFRI's, to ensure that the Dutch investments would be in line with European investments. This indeed happened; yet, another new committee was established to develop the first Dutch roadmaps. The committee established 11 evaluation criteria. Six were taken directly from ESFRI and five were specific to the Netherlands. The latter are interesting because they provided insight into the role of RIs in the Dutch science system. These five evaluation criteria were: (1) whether the RI provided a possible focus point for the Netherlands, (2) whether there was a critical mass of researchers in the Dutch science system, (3) whether there was sufficient institutional embedding, (4) whether there was a proven will to collaborate in the field and (5) whether the project aligned with broader societal developments (Commissie Nationale Roadmap Grootschalige Onderzoeksfaciliteiten 2008: 17–19).

Similar to NWO Investment Grant Large, this funding arrangement's evaluation criteria synced with the intention formulated in the science policy discourse outlined above. The Dutch roadmap for RIs was renewed in 2012 and again in 2016, and every few years (2012, 2014, 2018) NWO has organized funding invitations for projects included on the roadmap. Investments in RIs have thus significantly increased over the past 15 years, largely in line with European investments in RIs.

3.3 The Emergence and Institutionalization of Digital Humanities in the Netherlands

The rise of digital humanities as an interdisciplinary research area was preceded by the rise of computational methods in specific humanities disciplines. Arguably the most important, and most influential, was the development of computational methods in linguistics. The field of computational linguistics dates back to the 1950s (Van der Beek 2010) and many of its analytical tools are used in digital humanities research today. However, crucially, digital humanities is a broader research area than computational linguistics. Stemmatology, for instance, which "concerns itself with the problem of the genealogy of variant versions of manuscripts and print books" (van Zundert and van Dalen-Oskam 2014: 4) draws on philology and bio-informatics. Similarly, digital humanities research in art history and media studies employs a broader range of research methods.

Digital humanities emerged in "computer and humanities" groups in the second half of the 1980s at universities across the Netherlands. Most famous in that period was the Alfa-informatics lab at the University of Groningen, established in 1986 (van Zundert and van Dalen-Oskam 2014). This first wave of interdisciplinary "computer and humanities" groups, however, turned out to lead a fragile existence as they were often considered information technology helpdesks, and most were consequently defunded in the early 1990s when such help was considered no longer necessary (van Zundert and van Dalen-Oskam 2014). A few groups survived, in Groningen and Utrecht, and in the 1990s the KNAW became a more prominent actor in the development of digital humanities in the Netherlands.

The humanities institutes of the KNAW were, and are still, "collection-heavy", having had a function to develop an archive that predates their function as research organizations. Digitization initiatives in the 1990s were primarily focused on archives that were deemed crucial for cultural heritage. This is also why the Royal Library became a key actor in digital humanities in the same period, as well as the newly established Netherlands Institute for Sound and Vision. While digital humanities as a research area was already well established, its institutionalization in the KNAW as a research area only then, in the mid-1990s, became a topic of interest.

At the KNAW, both policymakers and researchers became increasingly aware throughout the 1990s what digitization would mean for collections and information management, as well as for research practices. The influence of digitization on research practices, and the new opportunities it would bring, was the subject of a report written in 1997 by the board of humanities of the KNAW. Two additional studies under the auspices of the AWTI, that reflect on digitization in the humanities and social sciences, were also released (Drosterij et al 2000; Bijker and Peperkamp 2002). The KNAW also invested in digital humanities research itself. At the end of the 1990s, the KNAW started the NIWI, an institute for scientific information that had a small but active digital humanities group called NERDI (van Zundert and van Dalen-Oskam 2014: 6). Out of the NERDI group, the Virtual Knowledge Studio (2005–10) emerged as an institute for what was then called "e-humanities". The Virtual Knowledge Studio was later replaced by the e-humanities group (2011–16), led by Sally Wyatt. This smaller research group sustained a network of scholars and projects working in digital humanities at the different KNAW institutes. In 2016, three humanities institutes of the KNAW intensified their collaboration under the name of the Humanities Cluster, and a new digital humanities group was formed led by Marieke van Erp. At the same time, pockets of digital humanities research continue to be active among the various KNAW institutes. Therefore, the KNAW was a crucial actor in the development of digital humanities in the Netherlands. In the next section, the role of the KNAW institutes will also become apparent, because, in one way or another, at least one KNAW institute participated in all RI projects in the humanities funded since the early 2000s.

4. THE DIGITAL RI FUNDING BOOM IN THE HUMANITIES

In the previous section, I told three short histories that serve as a backdrop for our understanding of the meaning and consequences for the sharp increase in funding for digital RI projects in the humanities after 2003. I detailed how the fragmentation and perceived lack of societal value of the humanities was an ongoing concern in science policy since the late 1970s. Furthermore, I traced the rise of instrumentation and RI funding arrangements, including their evaluation criteria. Lastly, I described the emergence and institutionalization of digital humanities as a research area, specifically the increasingly dominant position of the KNAW in this process. In this section, I first describe the funding boom that happened after 2003 in funding for digital RI projects in digital humanities. Second, I connect this development with the mainstreaming of digital humanities research more broadly and describe the promise that digital humanities research, specifically as envisioned to be delivered by CLARIAH, holds in the science policy discourse.

4.1 The Digital Research Infrastructure Funding Boom

The first humanities project that received funding through NWO Investment Grant Large was "Life Courses in Context" (€3.1 million), a project in socio-economic history, which started in 2003, only two years after the RAND Europe report. It was located primarily at the International Institute of Social History (a KNAW institute). The project extended the already existing project that established a historical sample of the population of the Netherlands, covering microdata of over 40 000 individuals, born between 1863 and 1922. It is important to note that this project had a much longer history that dated back to the 1980s, and had received funding from a variety of different sources. This digital RI, and the underlying research paradigm, precede this specific grant. There was already a tradition of using large quantitative databases, computation and statistical methods in socio-economic history. This is also true for other humanities fields that received infrastructure funding later on; there has always been a research tradition in place (Agar 2006; van der Beek 2010; van Zundert and van Dalen-Oskam 2014). What was new, however, was that this type of research was suddenly actively supported through a large infrastructural grant of the main Dutch funding organization.

"Life courses in context" was thus significant because it showed that humanities research could be funded through a funding arrangement traditionally reserved for engineering and natural sciences. It showed, moreover, that their request for infrastructural funding was institutionally legitimate. Of course, this RI was, at this point, of limited size. The sources that were digitalized covered a hallway in the institute; the data were distributed online on a simple website and were relatively small. In terms of collaboration, the data were used within different groups of socio-economic history as well as by historical sociologists and in historical research in health (for overviews see Kok et al 2009; Kok and Wouters 2013). But this project was just the start of a larger development in the Netherlands that coincided with the increasing focus on infrastructural funding on the European level as described above.

After the establishment of the ESFRI, the NWO-BIG supported a project to digitize a large collection of newspapers led by the Royal Library. After the first Dutch roadmap for RIs was developed, a subsequent funding call in 2008 led to funding for CLARIN-NL (€9 million); a digital RI in the field of computational linguistics.

In the second round of funding in 2012, a collaboration between CLARIN-NL and DARIAH-NL proposed to develop CLARIAH, a digital RI that would contribute to both the European CLARIN and DARIAH projects and develop a virtual research environment for the humanities. This project was spearheaded by three institutes of the KNAW, as well as Utrecht University, the University of Amsterdam, the Netherlands Institute for Sound and Vision, and included participation of all other Dutch universities and important archives and heritage institutions. The proposal was reviewed favorably and received €1 million in seed funding. In 2014, the revised CLARIAH proposal, which proposed to focus its efforts on the disciplines of linguistics, socio-economic history and media studies as frontrunners in (computation and) digitization of research, was fully funded (€12 million). A new proposal that extended the CLARIAH infrastructure to more humanities disciplines (called CLARIAH-PLUS) received funding in 2018 (€14 million).

The collaborative spirit of the different disciplinary communities (in particular the merger of CLARIN-NL and DARIAH-NL) was received with much agreement according to the policymakers at the Ministry of Education, Culture and Science I interviewed, and aligns with the evaluation criteria discussed above that were also used in the roadmap funding arrangement. CLARIN and CLARIAH showed that humanities scholars could cooperate across disciplinary and institutional boundaries.

Next to these large grants for CLARIN-NL, CLARIAH and CLARIAH-PLUS, a number of smaller projects have been funded. Through the above-mentioned NWO Investment Grant Large, five humanities projects were funded in just four funding rounds (between 2010 and 2018). In all of them, institutes of the KNAW were involved as well as other partners already active in CLARIN and CLARIAH. Moreover, all developed their digital research infrastructure in collaboration with either CLARIN or CLARIAH. Taalportaal (2010, €1.7 million), in linguistics, was developed in close collaboration with CLARIN. CLIO-INFRA (2010, €1.4 million), in socio-economic history, was initially part of the infrastructure bid of DARIAH that was unsuccessful, but went on to get funding through NWO Investment Grant Large in 2010 and was also further developed in CLARIAH. Nederlab (2012, €2 million), in linguistics and literary studies, was developed in close collaboration with – and even received additional funding from – both the CLARIN- and CLARIAH-consortia. Both Golden Agents (2016, €3.6 million), social and cultural history, and REPUBLIC (2018, €2.5 million), political history, are explicitly embedded in CLARIAH. The amounts indicated above are all based on the NWO contribution to the project which excludes in-kind or financial contributions of institutes themselves (NWO 2019).

Between 2003 and 2018, ten projects with a combined budget of €50 million were all funded with the aim to develop digital RIs in the humanities. The projects targeted frontrunners in digitization, notably linguistics and various subdisciplines of history. However, other disciplines such as media studies and literary studies are now also being targeted in these projects. Involved institutes include KNAW institutes, universities as well as archives and cultural heritage organizations. Crucially, different digital research infrastructure projects actively engage with existing, larger initiatives, notably CLARIN and especially CLARIAH, arguing that their work will build on or are compatible with these larger RIs.

4.2 Science Policy and the Promise of CLARIAH

It is not a coincidence that newer and smaller projects actively engage with the larger research infrastructure projects, notably with CLARIAH. From 2015 onwards, CLARIAH was positioned, by science policy stakeholders, as the main digital RI in the humanities in the Netherlands in relation to which new, smaller digital projects could and should be developed. This meant that new digital RIs, for instance, used the generic linked "open data" database structure developed in CLARIAH or borrowed particular elements from it, such as authentication and provenance middleware. Moreover, in relation to CLARIAH, digital humanities research was increasingly mainstreamed through other funding arrangements at NWO as well. For instance, digitization of sources was introduced as a legitimate part of any project funding application at NWO. This was crucial for many digital humanities projects proposals that would often need to do some work on digitization of sources. Also, a number of funding arrangements were launched for research project funding with a strong digital component. For instance, different social science and humanities funding arrangements of the E-Science Center were established, including a number of thematic programs for collaborations between academia and creative industries including CATCH, which focused on cultural heritage.

In a strategy document by the board for humanities of NWO, called "humanities aimed for the future" that outlined some of these changes, the mainstreaming of digital humanities was explained as well as the role of the CLARIAH infrastructure:

> Digital humanities is an undercurrent in all open instruments and thematic programs of the domain [...] The developments in digital humanities will, as said, be present in all instruments and programs and not in a single specific program. For NWO-Humanities it is of importance that in all these developments CLARIAH is taken as the basis for the infrastructural facilities in the humanities. (NWO 2015: 7/18)

While the domain as a whole is considered to be in crisis, a particular aspect of humanities research – building large digital RIs, and to an extent digital humanities research itself – is increasingly successful in obtaining funding. This development is enabled by the main funding agency in the Netherlands, as well as other important science policy actors, that actively shape science policy and funding arrangements to support digitization in the humanities. But why, then, do they do this?

Drawing on analysis of science policy documents since 2000, I argue that infrastructural projects, and specifically CLARIAH, are seen to offer the promise of increasing integration of a domain that is traditionally regarded as epistemically fragmented, as I have shown above.

The first effect is the revolutionary possibilities that these digital RIs (and the big data contained in them) will have to offer the sciences (e.g. Leonelli 2016). This element is clearly visible in the policy documents related to the humanities. They speak of a digital revolution happening in the humanities that will fundamentally alter how humanities research is organized, including its methodologies and theories. In the 2010–15 strategic vision of the KNAW, this is neatly summarized:

> The unique collections and datasources of the humanities institutes are the basis for the development of an advanced technological infrastructure ("computational humanities") that will lead to a methodological and theoretical renewal of research in the humanities [...] the foreseen methodological renewal will make it possible

> to discover in data sources, that are until now seen as (relatively) independent from each other, patterns and relations that will lead to significant new insights in the nature and the development of complex societal processes. (KNAW 2010: 23)

The unique character of this shift is highlighted in a number of policy documents including the 2008 report by Cohen (2008: 16), and the KNAW report on the future of the humanities (KNAW 2012), quoted directly below. The NWO report on large-scale RIs in the humanities (NWO 2013) even starts in its very first sentence highlighting the "fundamental change" that the introduction of the computer and digitization of sources brings to the humanities. The consequences include a change in methodologies and theories, as well as the organization of research, illustrated here: "The large-scale infrastructure and the costly, continuous maintenance of this all demand new competences and ask different requirements from investments and long-term financing. These developments ask to take stock and reflect on the work and work processes" (KNAW 2012: 12).

The coordination required to develop and maintain these types of costly digital RIs is seen to be delivered through CLARIAH. The NWO strategy document regarding the roadmap for RIs in the humanities explains: "CLARIAH and the collaborating initiatives within it guarantees a further development of an accessible large research infrastructure in which many small digital infrastructural facilities are brought together and new activities are executed in a coordinated way" (NWO 2013: 2).

A second effect is the inter- and trans-disciplinary nature of the digital humanities that will influence the humanities as a whole. Research becomes interdisciplinary because knowledge from the engineering and natural sciences are introduced in the humanities through its new involvement with large data sets. As the vision document describes:

> [T]here is increased interdisciplinarity. This is not merely a matter of collaboration between disciplines in the humanities, but especially the cross-pollination between humanities and informatics, between humanities and natural sciences and between humanities and social sciences. The use of knowledge, methods, and techniques from other scientific domains will provide the humanities with new insights. (NWO 2015: 7)

But research also becomes trans-disciplinary within the humanities as the connection of data sets gives scholars the opportunity, or forces them, to ask questions beyond their discipline. For instance, the linking of entirely different corpora (letters, artworks, state documents, etc.) or corpora from different periods will raise research questions that connect and transcend disciplines requiring socio-economic historians to work with art historians, linguists and political historians and develop new integrated theoretical frameworks. A description of CLARIAH summarizes this vision:

> CLARIAH (Common Lab Research Infrastructure for the Arts and Humanities) develops a digital infrastructure that combines large collections of data and software from different areas of the humanities. In this way humanities scholars – from historians, literary scholars, archeologists to linguists, speech technologists and media scholars – can study discipline-transcending questions about, for example, cultural and societal change. CLARIAH receives 12 million for the development of research instruments and training of researchers. This project is of major importance for the development of the humanities in the Netherlands; a digital revolution is going on that will drastically change how research is done. The potential societal impact is similarly large. (NWO 2015: 14)

A third effect is the increasing societal value of digital humanities research. The first quote from the KNAW 2010–15 strategy document connects the infrastructural development and its revolutionary consequences to the humanities that can give new insights in societal questions; it reads: "Because of this, the humanities within the Academy will be able to offer new directions for solutions of societal questions [...] regarding these questions the institutes intend, on the basis of excellent scientific research, to deliver a substantial contribution to the political and societal policy agenda" (KNAW 2010: 23). This, I would suggest, relates to the longer discussion on the value of the humanities (and the sciences in general) in and for society that was the topic of discussion in all the reports of the committees outlined above about the humanities in the 1990s and 2000s. The digital turn in the humanities is expected to alter how knowledge is produced, what type of knowledge is produced and, consequently, also the societal value of this knowledge.

5. GOVERNING THE HUMANITIES THROUGH RESEARCH INFRASTRUCTURE FUNDING

In this chapter, I conceptualized RI funding as a particular tool that science governance actors can use to steer research. I argue that funding arrangements can be regarded as part of particular science policy discourses and, in that capacity, enact particular normative frames of what "good science" is.

I show that in Dutch science policy since the late 1970s, two ideas regarding science are dominant. The first is that there should be extensive coordination of research agendas across disciplines so that a critical mass of researchers could arise around certain research areas. The second is that science should produce societal value, in particular by research and teaching that aligns with the demands of the labor market. The humanities did not fit well with any of these ideas; they were considered epistemically fragmented and not producing sufficient societal value. Committee after committee was put into place to develop plans in order to "solve" the problem of the humanities.

While the domain at large was thus considered to be in crisis from the late 1970s onward, it did remarkably well in acquiring funding to develop digital RIs. Since 2003, ten projects with a total budget of €50 million have been awarded in the humanities. In comparison to, for instance, the social sciences, this is a much larger part of the total budget for RI development than the humanities ever received. This raises the question why the humanities have been so successful. Based on an analysis of policy documents, I argue that digital RIs in the humanities, and the related emerging research area of digital humanities, are seen to foster collaboration and coordination in the humanities, and thus might solve the fragmentation the humanities are considered to suffer from. CLARIAH is in this regard considered as the core digital research infrastructure through which large-scale change of research practices in the humanities might occur.

While, in this way, I showed how two tools of science governance – introduction of normative frames of "good science" and that of funding arrangements – were put into place. I have not yet discussed the effects these interventions have had on research practices in the humanities. This is a crucial question, as I have outlined in the theoretical framework, and one that I can only give a preliminary answer to. Based on insights from science and technology studies, as well as preliminary empirical observations, there is reason to be cautious about any claims of technology-induced paradigm shifts (e.g. van Zundert 2012; Agar 2006).

Jon Agar (2006), in his study of computerization, showed that computers were predominantly introduced to reproduce already existing manual computational methods. In the Dutch humanities, we similarly see that fields that have a strong tradition of quantitative and computational methodologies, such as linguistics and socio-economic history, are also early adopters of the digitization of research practices. The introduction of digital research environments to search data sources seems to also be adopted on a larger scale, as it reproduces in a digital environment searching in archives and libraries (Kemman et al 2014). But does it go beyond that?

What we can intuit is that the investment of €50 million in digital RIs has, at the very least, strengthened particular epistemic communities within the humanities that already used computational methodologies. Moreover, digital humanities were also taken mainstream through other funding arrangements, which might have strengthened the same research communities. As funding success rates in the main "open competition" in the humanities are extremely low, this funding source might have enabled computationally minded scholars to continue their research lines while other, non-computationally minded scholars might have had to abandon their research, or academia, altogether.

A crucial empirical question for the future is the extent to which digital humanities can be understood as a coherent, scholarly community of digital research in the humanities. Infrastructural projects, as discussed here, often develop databases with easy-to-navigate graphical user interfaces that incorporate certain generic analytical tools (for instance for network analysis). These digital research environments will serve a particular subsection of humanists and allow for digital access to data sources, and possibly the use of digital analytical tools (when these tools can be used to answer the specific questions users have). A different subset of digital humanists, however, do not use generic tools but develop tools themselves, drawing on, for instance, computational linguistics and data science. These scholars usually use more advanced computational techniques for which coding skills are necessary and use software packages such as R and Python. An important empirical question becomes, then, to what extent different types of digital research practices in the humanities can be served within these digital RIs. This is an especially pertinent question in relation to the distribution of research funding across different types of scholars. If available funding is invested in large digital RIs, what is left for computational approaches that do not fit within these?

On the European level, combining the efforts of DARIAH-NL and CLARIN-NL into a single RI is gaining traction in other countries. German (CLARIAH-DE), Flemish (CLARIAH-BE) and Austrian (CLARIAH-AT) varieties exist or have recently received funding. As such, the rise of RIs, in relation to the mainstreaming of digital humanities, is not a Dutch affair. The integration of research efforts in Europe through, amongst others, the European Research Area, allowed for European-level science policy and research infrastructure development to happen (Cramer et al, ch 1 in this volume; Ulnicane, ch 4 in this volume). The example of CLARIAH shows that RIs in the humanities on a national level demonstrate isomorphic tendencies as national research communities follow and adapt to ongoing infrastructural developments.

ACKNOWLEDGMENTS

I greatly appreciate the comments I received from Karina van Dalen-Oskam, Guus Dix and Max Kemman on previous drafts of this chapter as well as the support of the editors of this volume.

REFERENCES

Agar J (2006) What difference did computers make? *Social Studies of Science* **36** (6): 869–907.

Benneworth P, M Gulbrandsen and E Hazelkorn (eds) (2016) *The impact and future of arts and humanities research*. Palgrave MacMillan.

Bijker W and B Peperkamp (eds) (2002) *Geëngageerde geesteswetenschappen. Perspectieven op cultuurveranderingen in een digitaliserend tijdperk*. AWTI.

Blume S and J Spaapen (1988) External assessment and "conditional financing" of research in Dutch universities. *Minerva* **26**: 1–30.

Cohen (Commissie Nationaal Plan Toekomst Geesteswetenschappen) (2008) *Duurzame geesteswetenschappen*. Amsterdam University Press.

Commissie Nationale Roadmap Grootschalige Onderzoeksfaciliteiten (2008) *Nederlandse Roadmap Grootschalige Onderzoeksfaciliteiten*.

de Boer H, J Enders and U Schimank (2007) On the way towards New Public Management? The governance of university systems in England, the Netherlands, Austria, and Germany. In D Jansen (ed) *New forms of governance in research organizations*. Springer, pp 137–52.

de Haan J (1994) *Research groups in Dutch sociology*. ICS.

Drosterij G, J van den Hoven, G-J Lokhorst, J de Mul and I van der Ploeg (2000) *Calculemus@human. Een voorverkenning naar de plaats van de geesteswetenschappen in de informatiesamenleving*. CFIC.

Elzinga A (1993) Science as the continuation of politics by other means. In T Brante, S Fuller and W Lynch (eds) *Controversial science: From content to contention*. SUNY Press, pp 127–52.

Franssen T, W Scholten, L Hessels and S de Rijcke (2018) The drawbacks of project funding for epistemic innovation: Comparing institutional affordances and constraints of different types of research funding. *Minerva* **56** (1): 11–33.

Fujimura J (1987) Constructing "do-able" problems in cancer research: Articulating alignment. *Social Studies of Science* **17**: 257–93.

Gerritsen A (2002) *Vensters op de Wereld. De studie van de zogenoemde "Kleine Letteren" in Nederland*. KNAW.

Gläser J (2019) How can governance change research content? Linking science policy studies to the sociology of science. In W Canzler, S Kuhlmann and D Simon (eds) *Handbook of science and public policy*. Edward Elgar Publishing, pp 419–47.

Gläser J and G Laudel (2016) Governing science: How science policy shapes research content. *European Journal of Sociology* **57** (1): 117–68.

Hallonsten O (2020) Research Infrastructures in Europe: The hype and the field. *European Review* **28** (4): 617–35.

Kemman M, M Kleppe and S Scagliola (2014) Just Google it. In C Mills, M Pidd and E Ward (eds) *Proceedings of the Digital Humanities Congress 2012*. Digital Humanities Institute.

KNAW (2010) *Voor de wetenschap. De akademie in de kennissamenleving. Strategische agenda 2010–2015*. Available at www.knaw.nl/nl/actueel/publicaties/voor-de-wetenschap (last accessed October 1, 2019).

KNAW (2012) *Contouren van een vernieuwings- en stimuleringsprogramma voor de geesteswetenschappelijke instituten van de KNAW*. Available at www.knaw.nl/nl/actueel/publicaties/de-contouren-van-een-vernieuwings-en-stimuleringsprogramma (last accessed October 1, 2019).

Kok J and P Wouters (2013) Virtual knowledge in family history: Visionary technologies, research dreams, and research agendas. In P Wouters, A Beaulieu, A Scharnhorst and S Wyatt (eds) *Virtual knowledge*. MIT Press, pp 219–50.

Kok J, K Mandemakers and H Bras (2009) Van geboortebank tot collaboratory. Een reflectie op twintig jaar dataverzameling en onderzoek met de HSN. *TSEG/Low Countries Journal of Social and Economic History* **6** (4): 3–36.
Laudel G (2017) How do national career systems promote or hinder the emergence of new research lines? *Minerva* **55** (3): 341–69.
Laudel G and J Bielick (2018) The emergence of individual research programs in the early career phase of academics. *Science, Technology, and Human Values* **43** (6): 972–1010.
Laudel G and J Gläser (2014) Beyond breakthrough research: Epistemic properties of research and their consequences for research funding. *Research Policy* **43** (7): 1204–16.
Laudel G and E Weyer (2014) Where have all the scientists gone? Building research profiles at Dutch universities and its consequences for research. In R Whitley and J Gläser (eds) *Organizational transformation and scientific change: The impact of institutional restructuring on universities and intellectual innovation.* Emerald Group, pp 111–40.
Law J and E Ruppert (2013) The social life of methods: Devices. *Journal of Cultural Economy* **6** (3): 229–40.
Leonelli S (2016) *Data-centric biology: A philosophical study*. University of Chicago Press.
NWO (1999) *NWO-programma's voor apparatuur 1999/2000.*
NWO (2013) *Strategisch advies: roadmap naar een grote infrastructuur voor de geesteswetenschappen.* Available at www.nwo.nl/documents/gw/beleid-en-strategie-grote-infrastructuur-nwo-geesteswetenschappen---september-2013 (last accessed October 1, 2019).
NWO (2015) *Toekomstgerichte geesteswetenschappen. Op weg naar een nieuwe balans.* Available at https://www.nwo.nl/documents/gw/gw---toekomstgerichte-geesteswetenschappen.-op-weg-naar-nieuwe-balans (last accessed October 1, 2019).
NWO (2019) Overzicht van toekenningen in het programma Investeringen NWO-groot. Available at www.nwo.nl/onderzoek-en-resultaten/programmas/investeringen+nwo-groot/toekenningen/ (last accessed July 26, 2019).
RAWB (1990a) *De geesteswetenschappen in Nederland: een overzicht.* Raad van advies voor het wetenschapsbeleid.
RAWB (1990b) *De toekomst van de geesteswetenschappen in Nederland.* Raad van advies voor het wetenschapsbeleid.
Rushforth A, T Franssen and S de Rijcke (2019) Portfolios of worth: Capitalizing on basic and clinical problems in biomedical research groups. *Science, Technology, and Human Values* **44** (2): 209–36.
Staal F (1991) *Baby Krishna*. Report of the Advisory Committee of Small Languages.
van den Besselaar P and E Horlings (2010) *Focus en massa in het wetenschappelijk onderzoek: de Nederlandse onderzoeksportfolio in internationaal perspectief.* Rathenau Instituut.
van der Beek L (2010) *Van rekenmachine tot taalautomaat.* Rijksuniversiteit Groningen.
van der Meulen B (2007) Interfering governance and emerging centres of control. In R Whitley and J Gläser (eds) *The changing governance of the sciences*. Springer, pp 191–203.
van Zundert J (2012) If you build it, will we come? Large scale digital infrastructures as a dead end for digital humanities. *Historical Social Research/Historische Sozialforschung* **37** (4): 165–86.

van Zundert J and K van Dalen-Oskam (2014) Digital humanities in the Netherlands. *H-Soz-Kult*. Available at www.hsozkult.de/debate/id/diskussionen-2396 (last accessed December 17, 2019).
Versleijen A (ed) (2007) *Dertig jaar publieke onderzoeksfinanciering in Nederland 1975–2005*. Rathenau Instituut.
Vonhoff H (1995) *Men weeft kaneel bij 't lood*. Final report of the Commission on the Future of the Humanities.
Whitley R (1984) *The intellectual and social organization of the sciences*. Clarendon Press.
Whitley R (2008) Universities as strategic actors: Limitations and variations. In L Engwall and D Weaire (eds) The university in the market. Portland Press, pp 22–37.
Whitley R, J Gläser and G Laudel (2018) The impact of changing funding and authority relationships on scientific innovations. *Minerva* **56** (1): 109–34.

8. The role of research infrastructures in innovation systems: The case of Swedish participation in the Halden Reactor Project

Olof Hallonsten, Hjalmar Eriksson and August Collsiöö

1. INTRODUCTION

Current research and innovation policy is largely *system-oriented*, viewing universities, corporations, governmental agencies, research institutes and other organizations as part of *innovation systems* that are nationally, technologically, sectorally or regionally defined (Lundvall 1992; Edquist 1997). Such systems have an overall purpose and function of achieving innovation for economic growth. The systemic nature of innovation means that the partaking organizational actors and their subunits and members innovate together, in more or less stable and durable constellations and networks, in harmony with rules and regulations, norm systems and culture, and all other important elements that go under the collective name "institutions" (Edquist 2004). Contemporary research and innovation policy is geared to the trimming of these systems, beyond mere funding of research and development (R&D), to enable all of their parts to contribute most efficiently to knowledge- and technology-based economic growth. But in spite of this dominance of the system approach to innovation, there is still a strong focus in research and innovation studies on *universities* and *firms*. With few exceptions, other very important entities and resources in innovation systems are either overlooked or studied as they were also either academic organizations or for-profit companies (Crow and Bozeman 1998: 7; Gulbrandsen 2011: 215). Some attention has been paid to funding agencies (e.g. Braun 1998), public/private research institutes (e.g. Gulbrandsen 2011; Arnold et al 1998; Hallonsten 2018) and governmental R&D labs (e.g. Crow and Bozeman 1998; Hallonsten and Heinze 2012). But judging from the empirical content of the greater body of literature on innovation and innovation systems, it seems their only important component organizations are universities and firms, and the only important processes for innovation are research, technology transfer and the commercialization of research results.

In this chapter, research infrastructures (non-capitalized, for the time being) are highlighted as a separate type of entity in innovation systems with a potentially very important role. The increased attention to Research Infrastructures (RIs) (capitalized) in recent policymaking and political rhetoric (see Cramer et al, ch 1 in this volume) has not been matched by similar interest in innovation studies. With the help of a deepened understanding of the theoretical foundations of the innovation systems perspective, namely classical sociological differentiation theory, research infrastructures can be viewed as *functionally differentiated* entities or organizations in national, regional, sectoral and global innovation systems. Such a view suggests that research infrastructures should be treated with attention to their function and specific characteristics, both in research and innovation studies and in research and innovation policy. To substantiate these claims, we use a case study of the role of the Halden Reactor Project (HRP) for Swedish nuclear energy R&D. The HRP is an international collaboration that uses a nuclear research reactor, in operation since 1959, and other infrastructural resources at the site in Halden outside Oslo in Norway, and is currently run and operated by the Norwegian Institute for Energy Technology (Institutt for Energiteknikk, IFE). The case study builds on an evaluation of the Swedish participation in the HRP in the years 2006–14, undertaken by the consultancy firm Oxford Research in the spring of 2016, on charge by the Swedish Radiation Safety Authority (Strålsäkerhetsmyndigheten, SSM) which is the Swedish signatory member of HRP. The analysis and results of the evaluation form the empirical core of this chapter, and build on documentation provided by the HRP (including reports, budgetary material, lists of bi- and multilateral contracts, statistics on staff and participants in research projects, program plans and the HRP project database) and interviews. The evaluation report clearly shows that the HRP is a crucial resource for the Swedish research and innovation system of nuclear energy-related R&D, and that its role and function is not comparable with academic research environments, R&D units of private enterprises or any other performing unit in an innovation system. In other words, there is convincing empirical evidence that the HRP occupies a functionally differentiated and specialized role in the Swedish (and Nordic) nuclear energy innovation system. On the basis of these results, it can generally be argued that research infrastructures should be treated as functionally differentiated and specialized entities of innovation systems, with the immediate consequence that they also should be evaluated in their own right, with attention to the function they fulfill, which is to support and sustain a broad range of activities that are part of innovation processes, on a longer term and with broader scope than the comparably very transient and temporary activities of much academic and industrial R&D. Since the empirical study was undertaken that forms the backbone of this chapter (in 2017), the board of directors of the IFE has decided to close the HRP, and it is currently under decommission. Since the study specifically concerns the time period 2006–14, however, any results of these recent events are not included in the analysis.

The next section provides a theoretical framing for the argument, and Section 3 presents the case and its history in brief, based on secondary literature. Section 4 describes Sweden's participation in HRP and, conversely, the current role of the HRP in the Swedish nuclear energy sector and its R&D activities, on the basis of analysis of the evaluation report. The chapter concludes with a discussion, based on the theoretical and empirical sections, that also points the way to some policy implications and suggestions for future research.

2. THEORETICAL FRAMING AND FOCUS OF THE STUDY

Since its original emergence in the 1980s, the systems approach to innovation has made a journey through academic and policy spheres and established itself in a nearly hegemonic role as the basic conceptual framework for innovation studies. The usefulness of the framework stems in part from the holistic acknowledgment of all actors, organizations, institutions and processes that contribute in any way to innovation processes and to the maintaining of a society and economy conducive to innovation, which in turn enables an inclusive view of the impact of the work of the actors and the system as a whole, and a possible route ahead for a deeper and better motivated understanding of how innovation occurs in heterogeneous systems and how different actors actually contribute to the system, and by extension, to innovation. Further, on the basis of conceptual developments in the past two to three decades, the approach allows the conceptualization of innovation systems as national, regional, sectorial or technology-specific, and thus with different independent variables. In this chapter and the study it reports on, the focus is on the Swedish nuclear energy-related innovation system which can be seen as both national and sectorial/technology-specific.

While innovation systems of various sorts and their various conduciveness for innovation have been investigated in many empirical studies, these have mostly focused on universities, firms and the interactions between them. They have thus largely neglected the variety of other actors that evidently are part of the systems and should, reasonably, also have importance for their functioning and overall performance. There is, in other words, too little understanding of specialization and differentiation within innovation systems, and too little attention paid to the roles of (organizational) actors outside of the private sector, the academic sector and government. In part, this is probably due to the popularity of concepts like the "entrepreneurial university" (Etzkowitz 1983; Clark 1998) and the "triple helix" framework (Etzkowitz and Leydesdorff 2000) which view universities as the motors of innovation systems and advocate a vast expansion of the missions of universities rather than specialization and differentiation of functions among different organizational actors within systems. The issue of evaluation of impact and output, continuously more pronounced in research and innovation policy and funding, is related to this. There are evident risks of an all-too-narrow focus on the measurable and quantifiable outputs of R&D activities – publications and patents – and the use of these measures to evaluate the performance of entities in the system that are only indirectly involved in its productivity by providing crucial resources for experimentation and the like, such as research infrastructures of various types. As pointed out in a previous study, in many cases RIs "do not produce any science themselves, their users do" (Hallonsten 2016b: 486). Unless the evaluation of the performance of research infrastructures is built on a deeper understanding of their role in the systems, and their contribution to systemic innovation processes, such evaluation will be misdirected and, worst case, make it appear as if an RI with a crucial role to play is unproductive or underperforming.

A reasonable approach to remedy these deficits is to explore the sociological foundations of the innovation systems approach. These date back to Auguste Comte, Herbert Spencer and Émile Durkheim, who all viewed differentiation as a key process in the modernization of society, that enables growing pluralism and complexity to be contained within an integrated whole. Society is seen as a system which is internally differentiated with respect to functions (Parsons et al 1953; Parsons and Smelser 1956/2010) or in terms of separate spheres or orders (Weber 1915/2009), which were later called institutions (Merton 1949/68). Functionalist differentiation theory was never a coherent theory framework, and its theorists made very disparate and sometimes contradictory contributions to its development, but its key idea is that society is unified in the pluralism of differentiated spheres or subsystems that are connected and interrelated (Alexander and Colomy 1990). Systems theorist Niklas Luhmann, one of the most distinct contributors to differentiation theory, separated three forms of differentiation: segmentation, stratification and functional differentiation. Segmentation differentiates society into equal subsystems (e.g. families, nation states); stratification differentiates society into a priori unequal subsystems (e.g. class and other hierarchies); whereas functional differentiation is the key feature of modernity that allows complex society to function as an integrated whole, with simultaneously independent and interdependent subsystems that correspond to vital functions such as economic development and growth, participation in collective decision making and correction of injustices (Luhmann 1995: 190, 2012: 86–7, 95). Functional differentiation thus describes *specialization* in a systems perspective, and has some connection to the concept of "division of labor" as first theorized by economists Adam Smith (1776/2012) and David Ricardo (1817/2004). Similar ideas appear in Plato's *Republic* and are probably as ancient as civilization itself. On the basis of the weight given to the division of labor by economists, and to functional differentiation by sociologists, the following postulation can be made: It is essentially meaningless to adopt a systems view on anything without including a theory of functional differentiation or division of labor, since the very idea of a system that strives towards an overarching goal requires an understanding of how the elements of the system are specialized, differentiated and interact.

Concerning innovation systems, the overarching goal is innovation, and the system consists of entities that contribute to innovation on the basis of distinct and separated functions. These functions can and ought to be identified and described in empirical studies. Certainly, innovation systems also maintain segmentary and stratificatory differentiation (e.g. the separation of regional authorities or universities with geographically delimited catchment areas; and the separation of universities of elite status and less prestigious ones), but in order to understand the performance of the system, it is the identification of different (organizational) actors in the system based on the identification of what they *actually do* that matters: Universities, firms, institutes, funding agencies with different missions, regulatory authorities and so on have different functions for the system as a whole. The task for empirical studies is to identify and describe these in order to enhance the understanding of the systems and how they function.

This has relevance also for the question of output/impact and how to measure it. Many of today's quantitatively oriented evaluations of performance and quality of R&D make use of only a few very standardized measures, among which publications and citations, spinoffs and patents, and growth of individual firms and the economy in general, are the most common. Simplicity and comparability are their merits, but they are also dangerously close to mere reproduction and amplification of inequalities that stem from previous stratification, such as basic financial conditions, legal frameworks and original missions (Hazelkorn 2011: 29–81; Dill 2009). Aside from the inadvertent effects on governance and organization of research and teaching activities that the excessive performance assessment exercises have (e.g. Münch 2014), studies have also shown that the impact of R&D is not as straightforward and simple to measure, and that quantitatively based indicators fail to capture the chains of effects that may end up in unexpected contexts and times (Jacobsson and Perez Vico 2010; Mazzucato 2015: 93–119). Specifically for RIs, it has been shown that the contribution to innovation processes is typically indirect – they do not produce any science themselves, but function as resources for researchers who are for the most part funded by other organizations (universities, institutes, etc.) and are thus *enablers* of scientific productivity (Hallonsten 2016b). This makes the common evaluation of their performance with the help of publication databases essentially flawed, and leads moreover to absurdities in terms of costs in relation to the measurable output (Hallonsten 2014).

In other words, it is necessary to adopt a broader view on effects and outcomes that can capture the indirect and seemingly hidden impact of the operation of RIs, based on a deep and empirically obtained understanding of their functionally differentiated role(s) in innovation systems. This is done with a stepwise analysis backwards that begins with direct impacts and present conditions, and continues with a successive expansion of the scope of investigations backwards in time and outwards including impact of wider reach. Documentation is reviewed in reverse chronological order, and the analysis is complemented by a series of interviews with actors successively more peripheral to the study object. This was the method for the study of the HRP that forms the empirical backbone of this chapter. Practical circumstances always set constraints, and also the most inclusive view on impact and meticulous qualitative investigation of effects outside of plain view will lead to overlooking some aspects. The inquiry was delimited to the years 2006–14 in order to keep expectations reasonable and sharpen the focus.

Some tentative ex ante classification is both useful and necessary. In the following, a separation is made between academic and industrial actors and their participation in various activities within the HRP and how they benefit from this participation. Several loosely defined categories are also used to identify different forms of outcomes and impact, such as competence building, the maintenance and development of a knowledge base, contract research, interpersonal and interorganizational networks and so on. Scholarly publication in peer-reviewed journals is relatively insignificant as an outcome from HRP research activities and is also not a central part of HRP strategy, and the small size and rather strong regulation of the nuclear energy sector also makes patenting, licensing and the formation of startups a negligible impact of the collaboration and its R&D programs. Therefore, classic quantitative measures like publication, citation and patent counts are inadequate for the task at hand, and outcome/impact has instead been investigated by the use of more qualitative, exploratory and narrative methods, combined with some data on Swedish participation and involvement in key *activities* for knowledge and competence building at the Halden site, rather than their quantifiable outcomes.

3. THE HALDEN REACTOR PROJECT

The HRP is an international collaboration of 19 countries (Belgium, Czech Republic, Denmark, Finland, France, Germany, Hungary, Japan, Kazakhstan, Norway, Russia, Slovakia, South Korea, Spain, Sweden, Switzerland, the United Arab Emirates, the United Kingdom and the United States) under the auspices of the Organisation for Economic Co-operation and Development (OECD). The collaboration facilitates the use of the instrumentation and experimental resources available at the Halden research reactor facility, located in the town of Halden, some 100 km south of Oslo, Norway. The facilities are owned by the Norwegian research institute IFE, which also conducts other energy-related research at the site, partially with the use of the same facilities as is used within the HRP, for example on commission by Norwegian industry. The HRP is a collaborative organization with memberships renewed every three years (IFE 2008: 6). Currently, over 100 organizations in the nuclear sector regularly use the facilities, including the reactor and fuel industry, utility companies, research institutes and academic organizations, and various nuclear licensing and regulatory agencies.

The activities at HRP make use of the Halden Boiling Water Reactor (HBWR), exclusively used for experimental purposes (i.e. not for power production), and the Halden Man-Machine Laboratory (HAMMLAB), a test bed for so-called Man-Technology-Organization (MTO) research, which is research in the borderland between technical and behavioral sciences that aims to improve safe operations and management of facilities with elevated risk, such as nuclear power plants. HAMMLAB enables such research in a physical control room environment as well as a virtual reality center and the nuclear reactor simulators Hammlab Boiling water reactor simulator and Fessenheim Research Simulator for Hammlab. In total, the HRP has little more than 200 employees, of which approximately 100–120 are scientists (university graduates) and 25 foreign experts on temporary stays ("secondments", see below) (IFE 2019). The activities are organized in joint programs, whose results are collected in program achievement reports and shared between all members, and bi- or multilateral (commercial) programs, whose output is owned by the participants of the specific program. Two parallel joint programs are run in three-year periods, one within nuclear fuel and materials tests for reactor cores, and one in MTO research. These are multidisciplinary in nature and consist of several smaller activities in research areas with different foci and scope, ranging from very specialized to broader scientific and industrial relevance and attracting partners from all member countries. Three technical expert committees, the Halden Program Groups (HPGs), within fuel, materials and MTO, oversee research activities. In addition to the triennial achievement reports from the joint programs, the HRP publishes work and status reports and acts as host of several conferences, workshops, meetings and summer schools. The Enlarged Halden Program Group (EHPG) meetings, usually held every 18 months, attract several hundred participants and are key events in the international nuclear energy sector. An important channel for knowledge dissemination and competence building among the participating countries is the so-called "secondee" system which enables HRP member organizations to send staff to participate in research and training for limited periods of time.

The IFE was launched in 1948 under the name Institutt for Atomenergi (Institute for Atomic Energy), with the mission to develop and build research reactors, at a time when Norway was still pursuing a domestic commercial nuclear energy program (IFE 2008: 3). In 1955 the institute began the design and construction of the nuclear research reactor that is today's HBWR. One year before its completion, a collaborative agreement with 11 other countries (Austria, Belgium, Denmark, France, Germany, Italy, Luxembourg, the Netherlands, Sweden, Switzerland, the United Kingdom) (Beere 2007: 12) under the auspices of the Organisation for European Economic Co-operation (OEEC, the predecessor of the OECD), was signed that established the HBWR as an international research reactor, and the HRP as an international collaboration (IFE 2008: 3). Since then, the HRP collaborative agreement has been renewed every three years, together with collective decisions among the members on the joint research programs for the coming three-year period.

The HBWR was first constructed as a test facility to advance nuclear power as an energy source for the participating countries, and therefore research activities were initially almost exclusively oriented to fundamental reactor technology and physics (Njølstad 1999: 147ff). In the 1960s, as commercially viable nuclear power technologies concepts became available elsewhere and each country did not have to develop its own program, the HRP research activities shifted towards studies of reactor performance, including fuel and materials testing as well as process supervision and control systems, a focus that remains to this day. The implementation of so-called high pressure loops in the late 1960s, enabling simulation of the conditions in commercial reactors, made the HRP a leading center globally for research on nuclear fuel and also opened up new opportunities for studying materials under irradiation (Njølstad 1999: 276–99). Norway continued to entertain plans for a domestic nuclear power industry up until the mid-1970s, but shelved them when the oil repositories in the North Sea were discovered and exploitation of these started. After the 1979 Three Mile Island nuclear accident in the United States and its dire repercussions for the already heated nuclear energy debate in Europe, the plans were finally discarded altogether, and the HRP remained entirely an infrastructure for research (IFE 2008: 16). At the same time, the Institute for

Atomic Energy was renamed IFE and expanded its focus to energy-related research more broadly (IFE 2008: 19).

In the 1980s, the productivity and high quality of operations and utilization of the HBWR, and a renewed concern for safety issues on an international stage, attracted a growing international interest in the HRP, including not least a mobilization by both IFE and the HRP in the area of simulation of reactor operation and of control systems and safety (IFE 2008: 20). Consequently, in 1983, HAMMLAB was created, which enabled the systematization of MTO research to the benefit of the international nuclear industry as well as the Norwegian domestic petroleum and arms industry (Øwre 2011: 19ff). It also marked the start of a broadening of the scope of MTO research into a wider area of human reliability studies. The HAMMLAB simulator was constructed on the basis of the design of control rooms in Swedish and Finnish nuclear power plants, which made the test facility especially interesting for partners in these two countries (Øwre 2011: 21). In the 1990s, fuel and materials research activities also continued to grow, but tendencies of a separation between the two main areas of R&D were also visible. The joint programs became increasingly focused on reliability and safety, and saw an increasing commitment from partner countries as well as the joining of several new ones, while the development and optimization of reactor technologies and solutions became more and more concentrated to bi- and multilateral commercial projects and programs (Njølstad 1999: 463–76). The 1990s also saw a renewed debate in Norway over the Halden reactor, with a key critical point when a proposal was made in the Norwegian Parliament in 1993 to close it, and as a result the IFE made attempts to strengthen the position of the HRP collaboration and secure its future by making its infrastructures more accessible to Norwegian industry, and to engage in the issue of nuclear safety in Russia and Eastern Europe (Beere 2007: 4). The strategy appears to have been successful: Between 1991 and 1996 the total value of IFE contracts from Norwegian industry more than doubled, and international interest in the HRP also increased, with new countries attaining membership (IFE 2008: 21).

Since the turn of the century, a major increase in both bi- and multilateral research contracts has been seen, and today roughly half of the research projects that use the reactor are part of the three-year research programs, with the other half being bi- and multilateral contracts. In the time period of the investigation of Swedish participation in the HRP (2006–14), there was significant continuity between the three rounds of joint programs within fuel and materials research and within MTO research, which also testifies to the role of the HRP as an institutionalized and reliable resource and platform for collaborative R&D that produced data and results with long-term relevance for its partners. Physical or material durability also has a role to play in this regard, as some experiments in the fuel and materials research area can last longer than three years and in some cases even span several program periods, due to very fundamental technical preconditions for studies of nuclear reactivity and decay.

The IFE remains a Norwegian research institute funded and operated by the government, and responsible for the operation of the HBWR as well as other infrastructures on the site in Halden. The HRP, in turn, is an international research collaboration governed jointly by its member countries through a board of managers which is the supreme governing body of the collaboration and to which each signatory member organization appoints one representative. The HRP contributes to the funding of operations of the HBWR and other facilities in Halden, but most of all, it forms the administrative structure and platform for their international utilization. The board of managers decides on the focus and direction of the joint research programs, and approves experiments to be run with the reactor infrastructure, also those undertaken on the basis of bilateral agreements between the IFE and a partner in a foreign country, but normally not utilization of the MTO infrastructure, to which access is less tightly regulated. Typically, foreign countries are expected to be members of HRP before access is granted to them through bilateral programs (IFE 2019).

For the following, it is necessary to note that the two research areas of the HRP are rather strictly separated in the physical and organizational structures of the site and collaboration, and connected to largely disjoint communities, as shown in Table 8.1. In the following, we build on previous analyses of large and complex research organizations that operate infrastructures for the use of multidisciplinary research communities. These studies have developed as an analytical framework a categorization of three different dimensions or components of these organizations, that correspond but that also display different characteristics and can be analyzed separately in order to achieve a fine-tuning of studies (Hallonsten and Heinze 2013, 2016; Hallonsten 2016a: 147ff). Table 8.1 cross-tabulates the two respective research areas of HRP with these three dimensions to explicate their distinctive features. The division is rather clear and straightforward but has some exceptions in the form of research groups that work across the divisions, and not least in the superstructure of HRP, which has a board of managers for the whole collaboration.

4. SWEDISH PARTICIPATION IN THE HRP

This section is in its entirety a summary of the evaluation report (Oxford Research 2016). Sweden has been a member of the HRP from its launch in 1958. The Swedish membership is currently organized in a consortium of six partners that all contribute to the membership fee, and with the SSM as coordinator and signatory member, and hence the body that appoints members to the Swedish seats in the HRP board and the HPG. In addition to SSM, the

Table 8.1 Organizational, infrastructural and scientific separation between the two HRP research areas

	Fuel and materials research	MTO
Organization	Fuel and materials division HPG Fuel HPG Materials	Safety and MTO Division HPG MTO
Infrastructure	HBWR	HAMMLAB Virtual Reality Center
Scientific fields	Physics Chemistry Engineering	Psychology Cognitive Science Information Technology

Table 8.2 Swedish membership fees in HRP, total 2006–14

	Total	Annual average
Total HRP budget (kNOK)	1 141 160	126 796
• Swedish contribution (kNOK)	55 700	6189
• Swedish contribution (%)	4.8	4.8
• paid by SSM (kNOK)	35 204	3912
• paid by industry (kNOK)	20 496	2277
• of which is in-kind (kNOK)	7670	852

Note: kNOK refers to thousands of Norwegian crowns.

consortium consists of the utility companies that operate the three Swedish running nuclear power plants (Forsmarks Kraftgrupp AB, Ringhals AB and Oskarshamnsverkens kraftgrupp OKG AB) and the two nuclear technology companies Vattenfall Nuclear Fuel AB and Westinghouse Electric Sweden AB. Currently, the director of the SSM research unit is the Swedish representative on the HRP board, and the HPG representatives are SSM staff in relevant and corresponding areas. Table 8.2 shows the membership fees paid by Sweden in the years studied here, and how these have been split between SSM and the Swedish industrial partners. While the Swedish share of the overall HRP budget looks small, the roughly 5 percent of the total that it represents is a significant share, given that several of the other 18 members in the HRP (see above) are much larger countries with much larger nuclear energy programs.

The five Swedish industry partners (see above) share their contributions to the Swedish membership fee almost equally. The two nuclear technology companies Vattenfall Nuclear Fuel and Westinghouse Electric Sweden contribute in-kind with fuel samples, and the utility companies contribute in-kind with operators for experiments in HAMMLAB.

Table 8.3 Total value of bi- and multilateral contract research within HRP with Swedish involvement, 2006–14

	Fuels and materials, 2009–14	MTO, 2006–14
Total value of contracts (kNOK)	587 300	454 000
Swedish share (kNOK)	35 400	47 100
Swedish share (%)	6.0	10.3

Note: kNOK refers to thousands of Norwegian crowns.

Table 8.3 contains details on the total value of bi- and multilateral research contracts between Swedish actors and HRP. In fuels and materials, the Swedish customers in the three studied program periods have been Westinghouse, Studsvik Nuclear, Ringhals AB and Chalmers University of Technology. In MTO, a number of different Swedish actors have been involved in the HRP on a contractual basis in the periods studied, for example the utility companies mentioned above, the mining company Luossavaara-Kiirunavaara Aktiebolag, the nuclear power safety and education provider Kärnkraftsäkerhet och Utbildning (KSU), Siemens Industrial Turbomachinery AB and the regional Swedish power production and distribution company Skellefteå Kraft. Looking at both Tables 8.2 and 8.3, we see that the relatively large Swedish share of the total HRP budget is to a significant part paid by industry, and this contribution is also matched by a rather large involvement in HRP research activities by Swedish actors on a shorter-term bilateral basis. Taken together, this is a strong indication that the HRP plays an important role in the Swedish nuclear energy industry.

A significant part of a country's benefit of membership in HRP is its ability to let individuals participate in different activities and programs in Halden, ranging from shorter events like workshops and summer schools to the longer-term "secondee" program, through which researchers, engineers and other professionals spend up to two years in Halden. In addition, researchers can spend limited periods of time at Halden as guest researchers and/or as part of their doctoral training, as part of the HRP collaboration. In addition, master students of Swedish universities regularly take part in research at the HRP, through the industrial partners where they do diploma work.

Several workshops are organized every year, and there are also recurring summer schools in each of the two research areas MTO and fuel and materials. Every 18 months, a two-day EHPG meeting is organized. Swedish participation in these activities is significant, with hundreds of attendees over the studied period (2006–14), and sometimes Swedes make up as much as one fifth of the participants in the summer schools, EHPG meetings and workshops. The majority of Swedish participants are affiliated with industry.

Table 8.4 Swedish involvement in HRP research as documented by HWRs, 2006–14

	Fuels and materials	MTO
Cooperation with Swedish university	2	1
Use of material from Swedish industry	17	
Use of reactors at Swedish nuclear power plants	9	
Swedish co-author	1	1
Swedish participation (operators)		10
Swedish participation (analysis)		3
Study on Swedish data and/or cases		10
Prototype development or similar with Swedish beneficiary		4
Other		4
Total number of HWR reports	*101*	*158*

The main channel for the documentation and dissemination of results and findings from research done within HRP are the Halden Working Reports (HWR), of which there are three main categories: (1) project result reports, the most common form of report, which document results from specific studies; (2) workshop documentation reports, the second most common, containing presentations and background material for specific workshops; and (3) literature reviews. The screening of HWRs undertaken as part of this study, summarized in Table 8.4, reveals a significant involvement by Swedish actors, of great variety and scope, but with a clear emphasis on industrial involvement which overshadows the rather limited academic participation.

The main industrial stakeholders in the fuel and materials area in Sweden are quite few, and they are both well known and rather tightly connected. In addition to the consortium partners (see above), one of the main industrial participants from Sweden is Studsvik Nuclear AB, a global R&D service consultancy group headquartered in Sweden. This company evolved from the former Swedish national research facility for nuclear research and research with high-intensity neutrons located in the small village of Studsvik some 70 km south of Stockholm, which ran its own research reactor until 2005. Since then, the contacts and relationship between the company Studsvik Nuclear AB and HRP/IFE have deepened significantly. As a main contractor for the government-controlled handling of nuclear waste (including irradiated fuel and materials samples) in Scandinavia, Studsvik has taken a leading role as a partner for HRP/IFE and for Swedish nuclear industry, and it is also the contractor for the transport of such samples to and from Halden, which also forms part of Sweden's in-kind contributions to HRP. SSM, as consortium partner and signatory member of HRP for Sweden, also has direct involvement in research activities within HRP, through its Department of Nuclear Power Plant Safety's Reactor Technology and Analysis Unit and Structural Integrity and Event Analysis Unit.

As can be deducted from Table 8.4, the degree of involvement in HRP by Swedish academics is relatively low. The recently signed Memorandum of Understanding between Uppsala University and HRP, which signals the intention of both actors to increase academic participation in research at Halden, is the first formal agreement between a Swedish university and the HRP/IFE. In addition to Uppsala University's Division for Applied Nuclear Physics, there has been some involvement from researchers at the Sustainable Nuclear Energy Centre at Chalmers University of Technology in Gothenburg, and the Swedish Centre for Nuclear Technology and Centre for Nuclear Energy Technology at the KTH Royal Institute of Technology in Stockholm. But interviewees in the academic sector report on the difficulties of getting the necessary insight into, and overview of, the foci and scope of the work within HRP for those not represented as consortium partners. Moreover, it seems the structure and composition of the collaboration is not particularly oriented to the academic community: Few HRP results are published openly, not even after the five years embargo is lifted, and the number of publications in academic journals that report on results from the HRP is indeed minor. Similarly, the EHPG meetings are seen as far less academically important and meritorious than other conferences in similar areas, which limits their relevance in this realm.

Only consortium members from Sweden participate in the planning of the successive three-year collaborative programs, which bars academic involvement but enables consensus around what Swedish interests should be guarded, since the people involved at the different companies and at SSM form a rather small and tightly knit network with long-standing and close collaborations and professional relations that have evolved both through joint R&D projects and through the regulatory and supervision role of the SSM in the Swedish nuclear energy sector. Participation in the HRP has, apparently, been a basis for the development of far-reaching cooperation in the sector which also has enabled the forming of R&D coordinating groups between and within the companies in the Swedish nuclear energy sector, to the benefit of the whole. According to interviewees, informal networks and personal relations have had a particularly important role to play: Strong interpersonal links have reportedly been established between individuals at key positions in the Swedish nuclear energy sector and HRP staff, that clearly go beyond the formal links established by Swedish (SSM) representation in the HRP board and the HPG.

The HRP brandishes some highly specialized capabilities in instrumentation and experiment design in the areas of nuclear fuel and materials research.

Views differ somewhat among the interviewees, but according to some, the capabilities to record real-time data from the reactor during operation are world-unique and especially useful for fuel tests, which is of crucial importance for the very application-oriented R&D undertaken by nuclear power plants in search of better safety and higher efficiency. Several interviewees have stressed that the participation of their company in HRP has secured crucial access for them to a research infrastructure that is both of a very high standard and conveniently accessible.

The most direct use of the HRP infrastructures by the Swedish nuclear sector is in the continuous refinement of nuclear fuel for Sweden's commercial reactors, and the processes for determining what specific fuel to use. Since in Sweden, the companies that operate the nuclear power plants are fully responsible for all aspects of safety of the plants, they have an obligation to verify and demonstrate the safety of the fuel they use, both when making modifications and when new or updated regulations or guidelines are implemented. This has led to direct use of the HRP results by these companies in their work of choosing nuclear fuel, and to the use of the HRP by fuel vendors in their work to support and compile safety documentation for their product. Such documentation typically builds on a wide range of supporting knowledge and data which can include results from different experiments and studies within the HRP joint programs, performed on a commercial basis at HRP, and of course also results from elsewhere.

The operators of Swedish commercial nuclear power plants also report on use of the HRP in for example demonstrations of safety of reactor technology, including the use of HRP data to inform models for the ageing and degradation of reactor construction materials and fuel behavior. Both Westinghouse Electric Sweden and SSM have used such data in their respective work to document reliability and safety.

On a general or indirect level, informants have testified to the role of HRP in sustaining a basic level of R&D capability in the Swedish nuclear fuel and materials area. The continuous testing and measurement at Halden, reported in HWRs, and documentation from the framework programs that are available to all collaboration partners immediately upon their release, continuously contribute to maintaining a high-level knowledge base among the Swedish consortium partners, and to push innovation. Informants from the industry partners and SSM alike have stressed this function. The relevance of the data and results produced within the HRP for the Swedish nuclear sector is secured by the frequent use of nuclear fuel and materials used in Swedish reactors.

On a related note, the role of the conferences, workshops and summer schools organized by HRP and IFE is also important for Swedish stakeholders. These platforms contribute to the dissemination of HRP results in an efficient manner and a very productive interactive setting, as well as more general knowledge of the fundamentals and developments within the fuels and materials area, and of course to the education and training of professionals in the sector. Interviewees also report on the contribution of these meetings to the forging of interpersonal and interorganizational relationships and the strengthening and maintenance of personal networks, and the rare opportunity they offer to gain insight into the agenda of the nuclear industry in other countries, which means that participation in HRP provides member organizations with a function of monitoring developments in the area on a global level.

In the MTO area, use and participation by Swedish actors is somewhat different, which is of course only to be expected given the differences between this and the fuel and materials area (see Table 8.1). The main Swedish users in the area are SSM and the operators of the Swedish nuclear power plants, who both have a direct interest in operations safety and reliability that, quite naturally, the nuclear fuel producers lack. In addition to the consortium partners, the nuclear power safety and education provider KSU, an independent company owned by Vattenfall, is a frequent participant in MTO research at Halden. In comparison with the fuel and materials area, the MTO research area of HRP has put greater effort in systematic coordination of commercial relations, including active work to match commercially relevant bi- and multilateral contract research with the different activities within the HRP MTO programs. This work has clearly benefited the relationship between Swedish stakeholders and activities in Halden, and has also strengthened interaction between Swedish companies, by contributing to the forming of important informal network ties. Several people from the operating staff at the Swedish nuclear power plants have also been hired by HRP as process experts through the years, and these individuals have reportedly contributed greatly to the HRP MTO research with their experience of the design and functionality of processes and systems in Swedish plants. Informal networks have thus grown strong in the MTO area just as in the fuels and materials area (see above), with many interpersonal contacts across organizations and sectors. The KSU is one clear beneficiary of this network building, but also connections and ties between Linköping University, Chalmers University of Technology, SSM and HRP have been formed on the basis of the many years of intense collaboration, in different shapes and forms, around MTO activities at Halden. Clearly, for Sweden, which has no corresponding infrastructure of its own, the role of the MTO research facilities in Halden has been significant, providing simulation infrastructures partly modeled on Swedish nuclear power plants, and in other ways stimulating the forming of important networks and providing a solid knowledge and competence base for industry and the SSM to draw on.

5. DISCUSSION AND CONCLUSIONS

The analysis in the previous section provides rather strong evidence, both quantitative (see Tables 8.2–8.4) and qualitative (interviews), that the HRP is very highly integrated in the Swedish nuclear energy innovation system and that especially the Swedish nuclear industry relies strongly on the HRP and the experimental resources and scientific infrastructure available at the Halden site, as well as the knowledge and competence base that has been built up and maintained on the basis of HRP collaboration.

Conceptually speaking, RIs take a functionally differentiated role as stable and durable resources and enablers of scientific research. They are, hence, not knowledge producers comparable with e.g. universities, university departments or governmental research institutes, and should not be viewed or evaluated as such. Both the physical realness and the resilience of the organizational arrangements (see D'Ippolito and Rüling, ch 11 in this volume; Cramer, ch 3 in this volume) put in place to operate e.g. particle accelerators, reactors, large telescopes, etc. make them enduring parts of innovation systems, and natural platforms both for breakthrough research of a momentary nature and institutional transformations of whole research and innovation systems (Hallonsten and Heinze 2013, 2016). The very practical/technological restriction on decommissioning times and procedures for nuclear reactors, applicable in the presently analyzed case, should also be considered in the analysis of the role it has taken in the Swedish (and global) nuclear energy innovation system. There is also a scientific/organizational logic to this: When RIs provide services of a very high standard to a community where access to such services is essential for short- and long-term productivity and quality of results, which clearly is the case in the technology-intensive nuclear safety research and related fields, it is highly likely that the RI in question develops a niche of its own and grows, over time, to be an inalienable part of the innovation system(s) it serves.

Therefore, considering the long history of the involvement of the Swedish nuclear energy industry in the HRP, the systemic role of the collaboration that emerges from the analysis in the previous section is unsurprising: Through its history and its several long-term collaborative processes, the HRP has become institutionalized as a durable and dependable partner for several organizations in Sweden (and elsewhere). While this reliability is clearly not enough in itself, but needs to be combined with a high and stable level of quality of operations, it forms a solid basis for collaborative activities where quality can be grown and cultivated. Evidence in the previous section suggests that the HRP has delivered also on this account. The scientific impacts, measured with common indicators such as publications and citations, have been relatively limited, and it is therefore fair to say that the role of the HRP in the Swedish nuclear energy innovation system is on the applied or strategic side rather than the basic (although such labels are ambiguous and should be used with care). But industry has benefited greatly from the HRP. The involvement of the companies of the nuclear industry in Sweden in the HRP is multifaceted and deep, and conversely, the role of the HRP is therefore crucial for their long-term knowledge and competence building and not least their technological and organizational renewal. The role of the HRP and its two research areas in fuel and materials and MTO is also significant as a solid and reliable knowledge deposit and competence base, and its function as the originator and sustainer of crucially important informal networks between people and organizations in industry, the regulation and supervision authority SSM and to some extent also the academic sector is also clear. Knowledge and technology development are doubtlessly important production factors in the technologically advanced and highly regulated nuclear energy sector, and the various results and findings from research within the HRP have clearly been turned into such productivity in Sweden. The large sums of money spent on Swedish companies on contract research in bi- or multilateral projects, in fact greatly exceeding the Swedish membership fee in the years studied (see Tables 8.2 and 8.3), testify to this. The numbers are matched by results from the qualitatively oriented analysis in the second half of Section 3, where it became clear that deep relationships have developed over time between the Swedish nuclear energy sector and the HRP. An interplay between the social/organizational and the technological contexts is seen in the forging of these deep relationships. For example, the modeling of the HAMMLAB simulator on Swedish (and Finnish) nuclear power plant control rooms is both a consequence of, and a catalyst for, the integration. Sure, this tight relationship, built on informal networks that have developed over a long time between industry and its very close supervision and control agency the SSM, cannot be completely ruled out as a partial cause for the limited involvement of Swedish academia in the HRP. The relative lack of academic involvement is testimony to a suboptimal feature of the system and the role of the HRP in the system, that might be interpreted as a negative side effect of the long-term development of a clear (functionally differentiated) niche of a research infrastructure in a particular innovation system, although this does not necessarily follow logically from the theoretical underpinnings of the analysis.

The chapter's key conclusion, that also points the way to further analyses with complementary perspectives and other case studies, is therefore that the functional differentiation between the infrastructure (HRP), its users (the Swedish nuclear energy sector), the broker of the relationship and funder of the infrastructure from the Swedish side (the Swedish government, represented by the SSM) is important to acknowledge in order to understand the system and not least how and why the system functions and produces innovation. The usefulness of the concept of functional differentiation for enhancing analyses that take a systems view on innovation has been demonstrated, and the concept has significant further potential given its solid and rich anchoring in classic sociological theory including vast theorizing and empirical investigations that have contributed to its refinement in different aspects. Importantly, the concept also allows and calls for a necessary shift of focus in performance evaluation, away from the simple indicators of patent counts, bibliometrics and spinoffs/foundations of companies, which are largely unworkable in this context, as the analysis has shown. Research designs must therefore be developed that go beyond the easily quantifiable and use document analysis, interviews and reiterative analyses of materials to establish a broader and deeper qualitative view on the impact and roles, also on the micro and meso levels. This will lead to a better understanding of how actors and entities in innovation systems are functionally differentiated, which will be a useful insight for innovation policymakers, not least when it comes to research infrastructures whose alleged importance in innovation systems keeps growing (see Cramer et al, ch 1 in this volume; Bolliger and Griffiths, ch 5 in this volume).

Regarding policy implications, a pedagogical way of arriving at these is to pose the question of what alternative investment the Swedish nuclear sector and the regulatory agency SSM could have made instead of its involvement in the HRP. The question is not only counterfactual but also preposterous since it supposes that SSM and the nuclear sector can act as a monolith and make collective strategic decisions for themselves, along these lines. But this is exactly the point: The HRP is a resource and a component in the system where SSM and the Swedish nuclear industry companies are active and which they jointly sustain and innovate through, and the HRP has a role in this system that makes it impossible to replace. Infrastructure investments like Swedish membership in the HRP should thus be treated differently than other investments as part of innovation policy, and the infrastructures should be acknowledged for their functional specialization so that more fine-tuned policymaking can be achieved and more adequate priorities made.

The generalized conclusion is that in spite of the impression given by some literature on the topic, innovation systems consist not only of academic research environments with some emphasis (but no exclusive focus) on fundamental or curiosity-driven research, companies and their R&D units where emphasis lies on applied research and product/process development, and regulating actors such as the state and other authorities that set the framework. Functionally, innovation systems consist not only of research, product and process development, commercialization and marketing and education. Other actors and processes may well be just as important, and functional differentiation between them is likely only reinforced and accentuated by developments that increase complexity, such as technological "sophistication" and disciplinary specialization and reorganization (e.g. Ziman 1994). The major contribution of this chapter has been the identification of the role of infrastructures as resources but also as sustainers of some very important processes and maintainers of knowledge, competences and important interpersonal and interorganizational networks. These functions, and the entities that have these functions, deserve more attention.

ACKNOWLEDGMENTS

The evaluation that forms the empirical core of this chapter was carried out on charge by the Swedish Radiation Safety Authority (Strålsäkerhetsmyndigheten, SSM).

The work to finalize the chapter was funded by the Swedish Research Council, grant no 421-2012-519.

REFERENCES

Alexander J C and P Colomy (eds) (1990) *Differentiation theory and social change: Comparative and historical perspectives*. Columbia University Press.

Arnold E, H Rush, J Bessant and M Hobday (1998) Strategic planning in research and technology institutes. *R&D Management* **28** (2): 89–100.

Beere W (2007) *The Halden Reactor Project: Experience gained in international research*. Paper presented at the IAEA International Conference on Research Reactors: Safe Management and Effective Utilization, November 5–9, Sydney.

Braun D (1998) The role of funding agencies in the cognitive development of science. *Research Policy* **27**: 807–21.

Clark B (1998) *Creating entrepreneurial universities: Organizational pathways of transformation*. IAU Press.

Crow M and B Bozeman (1998) *Limited by design: R&D laboratories in the U.S. national innovation system*. Columbia University Press.

Dill D D (2009) Convergence and diversity: The role and influence of university rankings. In B M Kehm and B Stensaker (eds) *University rankings, diversity, and the new landscape of higher education*. Sense Publishers.

Edquist C (1997) (ed) *National systems of innovation*. Pinter.

Edquist C (2004) Systems of innovation: Perspectives and challenges. In J Fagerberg, D Mowery and R Nelson (eds) *The Oxford handbook of innovation*. Oxford University Press.

Etzkowitz H (1983) Entrepreneurial scientists and entrepreneurial universities in American academic science. *Minerva* **21** (2–3): 198–233.

Etzkowitz H and L Leydesdorff (2000) The dynamics of innovation: From national systems and "Mode 2" to a Triple Helix of university-industry-government relations. *Research Policy* **29**: 109–23.

Gulbrandsen M (2011) Research institutes as hybrid organizations: Central challenges to their legitimacy. *Policy Science* **44**: 215–30.

Hallonsten O (2014) How expensive is Big Science? Consequences of using simple publication counts in performance assessment of large scientific facilities. *Scientometrics* **100** (2): 483–96.

Hallonsten O (2016a) *Big Science transformed: Science, politics and organization in Europe and the United States*. Palgrave Macmillan.
Hallonsten O (2016b) Use and productivity of contemporary, multidisciplinary Big Science. *Research Evaluation* **25** (4): 486–95.
Hallonsten O (2018) Development and transformation of the third sector of R&D in Sweden, 1942–2017. *Science and Public Policy* **45** (5): 634–44.
Hallonsten O and T Heinze (2012) Institutional persistence through gradual adaptation: Analysis of national laboratories in the USA and Germany. *Science and Public Policy* **39** (4): 450–63.
Hallonsten O and T Heinze (2013) From particle physics to photon science: Multidimensional and multilevel renewal at DESY and SLAC. *Science and Public Policy* **40** (5): 591–603.
Hallonsten O and T Heinze (2016) "Preservation of the laboratory is not a mission": Gradual organizational renewal in national laboratories in Germany and the United States. In T Heinze (ed) *Innovation in science and organizational renewal: Historical and sociological perspectives*. Palgrave Macmillan.
Hazelkorn E (2011) *Rankings and the reshaping of higher education: The battle for world- class excellence*. Palgrave Macmillan.
IFE (2008) *50 years of safety-related research: The Halden project 1958–2008*. Booklet, Institutt for energiteknikk – Halden Reactor Project.
IFE (2019) *The Halden Reactor Project*. Available at https://ife.no/en/project/the-halden-reactor-project/ (last accessed September 3, 2019).
Jacobsson S and E Perez Vico (2010) Towards a systemic framework for capturing and explaining the effects of academic R&D. *Technology Analysis and Strategic Management* **22**: 765–87.
Luhmann N (1995) *Social systems*. Stanford University Press.
Luhmann N (2012) *Theory of society*, Vol. 1. Stanford University Press.
Lundvall B Å (ed) (1992) *National systems of innovation*. Anthem Press.
Mazzucato M (2015) *The entrepreneurial state: Debunking public vs. private sector myths*. Public Affairs.
Merton R K (1949/68) *Social theory and social structure*. Free Press.
Münch R (2014) *Academic capitalism: Universities in the global struggle for excellence*. Routledge.
Njølstad O (1999) *Strålende Forskning: Institutt for Energiteknikk, 1948–1998*. Tano Aschehoug.
Øwre F (2011) The history of HAMMLAB: 25 years of simulator based studies. In A B Skjerve and A Bye (eds) *Simulator-based human factors studies across 25 years: The history of the Halden Man-Machine Laboratory*. Springer.
Oxford Research (2016) *Evaluation of the Swedish participation in the Halden Reactor Project 2006–2014*. Report.
Parsons T and N J Smelser (1956/2010) *Economy and society*. Routledge.
Parsons T, R F Bales and E Shils (1953) *Working papers in the theory of action*. Free Press.
Ricardo D (1817/2004) *On the principles of political economy and taxation*. Dover.
Smith A (1776/2012) *An inquiry into the nature and causes of the wealth of nations*. Wordsworth.
Weber M (1915/2009) Religious rejections of the world and their directions. In H Gerth and C W Mills (eds) *From Max Weber*. Routledge, pp 323–59.
Ziman J (1994) *Prometheus bound: Science in a dynamic steady state*. Cambridge University Press.

9. The access and return on investment dilemma in Big Science Research Infrastructures: A case study in astronomy

Andrew Williams and Jean-Christophe Mauduit

1. INTRODUCTION

Big Science, as described in this book, is science conducted on big machines operated by big organizations surrounded by big politics (Cramer et al, ch 1 in this volume). This conceptualization of Big Science incorporates the notion of a "transformation". In the decades immediately after World War II, Big Science was characterized by large facilities constructed and operated by relatively closed scientific communities for ostensibly scientific reasons, but wrapped in geopolitical and military factors and the superpower rivalry of the era. The transformation of Big Science refers to the gradual change to large facilities opening up to wide, diverse scientific and industrial user communities coupled with the emergence of contemporary imperatives to demonstrate increased and broader relevance to society and public policy (Hallonsten 2016) – one might say, the need to have "Big Impact".

While scholarly research has examined the effects of this transformation in terms of the history, politics, funding, governance, organization and sociology of Big Science, the topic of who gets access to facilities as well as how and why this access is allocated remains understudied. In practical terms, the design and implementation of access policies for Research Infrastructure (RI) projects have been developed on a case-by-case basis, seemingly without reference to a set of principles or framework. The access question is fundamental for both the scholarly study of Big Science and the constellation of actors that support, fund and lead big machines and big organizations for research. From the perspective of scientific users, the importance of gaining access to experimental or observation time on large RIs is approaching that of grant awards and publications in terms of establishing individual and institutional productivity and credibility (Hallonsten 2016). From the political perspective, policymakers and funders face contradictory policy pressures. On the one hand, funders need to demonstrate good value for money, leading to a focus on industrial uses or technology transfer, and return on investment or guaranteed access for the funder's scientific community, so-called "fair return", or "juste retour" (Hallonsten 2014b: 43). On the other hand, facilities are encouraged to be as open and as internationally oriented as possible, for scientific but also political reasons (Cramer, ch 3 in this volume). For any given RI, the maximization of its scientific outputs in terms of excellence necessarily requires excellence to be the only selection criteria for access, rather than national or institutional affiliation. This is in contradiction, however, to many other policy requirements to ensure "juste retour".

This chapter explores these dilemmas in the era of transformed Big Science and thus contributes to a richer understanding of the various factors that make it challenging and complex for policymakers and leadership to define who should get access, under what circumstances and for what reasons. The chapter is structured as follows. First, several broad policy themes that frame and provide context to the overall question of access to Big Science facilities are described. Second, we explore the issue of access using astronomy as an example, by reviewing the processes, tensions and history of access to astronomical observatories. Third, using an economic goods approach as a lens, we explore the concepts that underpin two common justifications held by policymakers and scientists alike for understanding access in relation to "return on investment": firstly, that returns should accrue primarily to individual scientists, and to the institutions or communities that fund facilities; and secondly, that access to facilities represents a broader public good. Finally, conclusions and policy implications are offered, which point towards further research needed on this topic. At the time of writing, the question of access to research infrastructures has not been independently addressed in the scholarly literature, which makes this chapter an important contribution, both in terms of its scholarly relevance and its potential usefulness for policymaking. While the empirical focus of the chapter is astronomy, the issues are equally as relevant to all user facilities, including synchrotron radiation, free-electron laser and neutron beam facilities.

2. PLACING THE ACCESS QUESTION IN CONTEXT

Several broad trends in the landscape of transformed Big Science contribute importantly to the context of the question of access to RIs, and highlight its significance. These trends show no sign of changing course in the near future and will continue to generate tensions in choices made by policymakers, funding agencies and scientists regarding participation in RIs and the respective return on investment.

The post-World War II era of Big Science witnessed growth in both publicly and privately funded RIs coupled with increasing internationalization (Cruz-Castro et al 2015), specialization and sophistication (Ziman 1994). According to some commentators, we are entering a "fourth age of research" dominated by "international collaborations between elite research groups" (Adams 2013: 557). Public funding, which was in earlier years concentrated in universities and some national laboratories, has transferred increasingly to highly specialized regional and international facilities (Thomasson and Carlile 2017). Particularly in the United States (US), continuing growth in private foundation funding, often with substantial endowments, has contributed to a proliferation of multipartner, large-scale projects (NSF 2018; Grant 2017). The focus on scientific excellence, publication and grant performance, coupled with generally strong funding commitment from governments, has led to intense competition to access top-performing, cutting-edge research infrastructures in order to conduct science or be involved in their design, construction and operation.

These forces of growth and competition, operating at both individual, university and national levels, have ensured that cutting-edge science increasingly requires large facilities and high budgets, creating a *fait accompli* towards multipartner and international arrangements. In order to be scientifically competitive, many RIs *have* to be international or multipartner to achieve the scale of financing and the scientific or technical expertise required. This situation is particularly apparent in Europe, where the many smaller European scientific communities and their respective countries' science budgets cannot compete internationally without joining forces and pooling resources. The long list of multi- and international partner research infrastructures on the European Strategy Forum on Research Infrastructures (ESFRI) roadmap (Bolliger and Griffiths, ch 5 in this volume) bears testament to this trend. Furthermore, to be eligible for many categories of research infrastructure funding attributed by the European Commission's framework programs for research and innovation, proposals are required to be composed of partners from three or more countries. This multipartner and multinational nature creates complexities in how the benefits of participation, namely access to the facility, is distributed between members but also non-members that have not contributed to the funding or design of the facility.

Another normative theme emerging in the RI policy domain is that facilities, particularly those that are publicly funded, should evaluate their scientific productivity and also their socio-economic impact beyond direct scientific achievements (Giffoni et al 2018; Hallonsten 2016). The OECD has engaged in an extensive exercise to create policy guidance and lists of metrics to conduct this task. The European Competitiveness Council of Ministers recently gave a mandate to the ESFRI to develop a common approach for the monitoring of their performance (EU COMPET 2018). While the concept of socio-economic impact assessment is certainly not new, the increasing requirement for research infrastructures to conduct these assessments speaks to the increasing cost, size and political significance of the facilities. Aside from their scientific goals, large facilities inevitably become a matter of regional development, industry and innovation policy, where the high use of public funds necessarily creates a need for an evaluation of policy impacts (Griniece et al 2015). An important question surrounding the various imperatives to measure impact and productivity and the associated need for funders to demonstrate greater uses beyond fundamental science is the extent to which they affect the access and selection criteria of scientific projects for a facility.

These three trends of internationalization, competition and impact create a series of dilemmas for any given user facility. The increasing internationalization and multipartner nature of large facilities generates substantial complexity over how financial contributions relate to access to use of the facility, or to other benefits such as industry contracts and governance involvement. The increasing competitiveness and demand for "excellence", and their links to the rewards and credibility of individual scientists, institutions and facilities, increase the pressure to gain usage of top research facilities. The need to demonstrate "Big Impact" and increase societal relevance implies a move away from excellence-driven missions and the accommodation of diverse industry uses and devoting resources away from science operations to technology transfer or public outreach.

3. THE ACCESS QUESTION IN ASTRONOMY

The trends of internationalization, competition and impact outlined above generate tensions and issues in determining who gets access to research infrastructures. For several reasons, astronomy is an instructive example of these various issues.

First, astronomy displays elements of both continuity and change in terms of its transformation. Astronomical sciences are characterized by a mixed environment of large facilities open to a wide range of users, but also others that are very restricted. The user community is highly international (Coccia and Wang 2016) and dispersed with a strong differentiation between user and operator (Hallonsten 2016), but highly segmented in certain respects between subfields.

Second, astronomy is in a worse case, relative to other fields, with respect to being able to adapt to the various policy contradictions highlighted in the previous section. In comparison to synchrotron and neutron beam facilities, which can "sell" their services to industry and conduct "policy-friendly" activities such as industrial partnerships and technology spinoffs, astronomy is indeed quite limited in this regard. While substantial industry contributions and innovation occur during construction phases of large and technologically complex astronomy facilities, the possibilities are much smaller during the operational phases. Without the additional policy benefits of industrial partnerships and spinoff companies in socially relevant fields, such as health technology and medicine, astronomy facilities therefore fall on the margins of the "innovation system" in the operations phase (Hallonsten 2016).

Moreover, the "users" of the facility only hail from academia and primarily seek to answer fundamental science questions in astronomy, contrary to facilities in other fields which serve a diverse and multidisciplinary user base (Hallonsten 2016). Funders therefore need to justify their investments on the pure value of fundamental science and scientific leadership for their own nations' or institutes' scientists. Having privileged access to the best instruments to do cutting-edge scientific research strengthens scientists' abilities to publish in top-ranked journals, gain tenure track or group leadership and attract subsequent grant funding. This, in turn, encourages the restrictive access policies in place in many of the astronomy research infrastructures worldwide. This policy dynamic, however, is generally at odds with the ideals of pursuit of science for the common good, and clashes with the stated values of the astronomy community.

3.1 Inside the Astronomy Communities: Research Conditions and Principles for Access

The field of astronomy has several key features that link scientific excellence strongly to the ability of scientists to freely access global facilities. First, with only about 13 000 professional astronomers registered with the International Astronomical Union (IAU) – one of the oldest scientific unions with a membership of professional astronomers from around the world – the field is relatively small in comparison to other scientific disciplines and is highly international by nature (Wagner et al 2016). Second, in contrast to this international user base, facilities are location-dependent and therefore fairly geographically concentrated, given the specific atmospheric conditions needed to observe the heavens, thus falling within the remit of certain nations or funding consortia. For example, isolated mountain tops with stable weather conditions provide optimal observing sites for optical telescopes; vast desert expanses are best for radio telescopes; and infrared or ultraviolet wavelengths need to be observed from satellites in orbit due to their absorption by the Earth's atmosphere. Third, while many facilities are often dedicated to observing a particular astronomical phenomenon or restricted to a specific range of the electromagnetic spectrum, cutting-edge astronomical research increasingly requires multiwavelength, multimessenger (from cosmic rays to neutrinos or even now gravitational waves) observations at once (Finkbeiner 2018). Hence, excellence in astronomy not only requires access to specific facilities which are located around the world, but also depends more and more on simultaneous access to different facilities, which may be operated by different funding consortia.

In addition to these physical characteristics and constraints particular to this field, as an epistemic community (Haas 1992), astronomy has developed, over the years, a set of shared norms that increasingly clashes with the demand for return on investment by funders. In that regard, the history of open access in astronomy in the US in the twentieth century can be particularly enlightening. In the pre-World War II era of astronomy, private foundations funded telescopes that benefited only a subset of American astronomers affiliated with those institutions that received funding. In the wake of the creation of the National Science Foundation (NSF) in 1950, this situation partly led to the advent of federally funded observing facilities such as the National Optical Astronomical Observatories (since October 1, 2019 called the National Optical-Infrared Astronomy Research Laboratory) and the National Radio Astronomy Observatory that guaranteed all astronomers in the US access to these top publicly funded telescopes. In turn, these facilities opened their time to astronomers worldwide, pushed by American radio-astronomers advocating for fairness and equity of access for all. This became known as open skies policy (OSP), and still applies to all national telescopes, even the most expensive ones like the National Aeronautics and Space Administration (NASA) Hubble Space Telescope. The principle was reiterated in the 2014 NSF Report of the Astronomy and Astrophysics Advisory Committee of the US. In the report, astronomers emphasized, among other points, that "the primary goal of the astrophysics community is to produce the best understanding of our universe, [… that] opportunities to participate in the implementing consortium of an astrophysics project or facility should be […] regardless of institutional or national affiliation, [and that] access to a large astrophysics project or facility (typically observing time) should be allocated through an open, merit-based process" (NSF 2014: 5). This apparent altruism of the US scientific community, however, may also be entangled with motives of self-interest. McCray (2000: 685) offers an excellent insight into the "moral economy" of astronomy and the post-World War II fight between the "haves" (privately funded telescopes and their community of privileged astronomers) and the "have nots", warning that "ideals of equity and fairness with respect to telescope access" can also be "used to serve broader institutional needs" (McCray 2000: 706). Similarly, astronomy RIs are inevitably embedded within a certain geopolitical background. In some cases, the rationale for large astronomy projects becomes intertwined with larger political narratives, such as national development or foreign policy objectives (Mauduit 2017; Pandor 2012; Pozza 2015). Nonetheless, the Astronomy and Astrophysics Advisory Committee (AAAC) report echoes the moral principles held by the astronomy community at large.

Turning to the IAU itself, as the official voice of the community, similar values are indeed found enshrined in its most recent strategic plan, where one can read that "the quest to examine the evolution of the Universe is an endeavour that unites all countries, for we are bound by the common goal of making sense of our place on Earth" (IAU 2018: 7). As an epistemic community, astronomers tend to think beyond national, political or economic borders and some of the most recent ventures of the IAU, such as the creation of an IAU Office of Astronomy for Development, further highlight this philosophy: its strategic plan states that "many large international telescope facilities are accessible to all astronomers throughout the world, providing an inexpensive entry to cutting-edge international research for developing countries" (IAU 2012: 11). Pushing further, "the long-term vision of the IAU is that [...] all countries will participate at some level in international astronomical research" (IAU 2012: 12), facilitating "the advancement of the next generation of astronomers and scientists" and stimulating "global development through the use of astronomy" (IAU 2018: 4). In summary, the collective global community embodied by the IAU understands astronomy research as an international endeavor uniting all countries, which should be available to researchers of any nationality and contribute to international development.

Yet the research needs and shared norms of the astronomy community are at odds with the identified consequences stemming from the need to finance, build and operate ever larger, more powerful telescopes coupled with little perspective of "return on investment" for the funding consortia in the operations phase beyond the publishing prestige of new discoveries contributing to advancing humanity's knowledge of the Universe. Indeed, in practice, astronomy access policies are quite heterogeneous and do not reflect these shared norms.

3.2 Astronomy Access Policies in Practice

In theory, the science-driven and egalitarian perspective is maintained by the use of Telescope Allocation Committees (TAC) that evaluate the scientific merits of the proposals and rank them in order of priority, for the best science to take place. The best scientific ideas should therefore take precedence and be allowed to be pursued and tested at the best possible facilities. Nationality or any other criteria should have no bearing on the science that is carried out. In practice, however, the situation can be different. Many telescopes have restrictive quotas for users from outside of their funding partner countries and while these may get to be reviewed by a TAC, usually only a fraction get awarded time, even if their proposals may be of equal or higher scientific merit. Unfortunately, many of these committees do not make public their acceptance rate statistics. This renders a comprehensive analysis of access policies difficult.

Nevertheless, even a limited sample is enough to bear witness to the diversity of these access policies. These fall onto a broad spectrum, from a few fully open (open time fraction of 1) to many in between or fully closed (open time fraction of 0). Some of the current access policies of specific international astronomical organizations and national telescope facilities are described in more detail in Mauduit (2017). On the high end of the spectrum, one will find the fully open telescopes, such as India's Giant Metrewave Telescope or the US's Hubble Space Telescope, which accept all international proposals and publish acceptance rate statistics in a transparent manner. Yet these are exceptions rather than the norm. Many domestic facilities tend to serve only their national astronomy communities or a very limited number of partner institutions, in line with the "prestige" return on investment argument: telescopes such as the Gran Telescopio Canarias or the South African Large Telescope are at the very low end of the spectrum, with a near zero open time fraction. Similarly, many multinational astronomy research infrastructures would also be near to fully closed if it were not for those in which the US is a partner through its national agencies, simply because of OSP applying to its observing time fraction. For example, the US NSF's share in the Gemini telescopes accounts for the large open access fraction on each of Gemini's two 8 m telescopes. Most other international facilities then fall towards the low end of the spectrum, with open shares of a few percent only or zero, a trend continuing with the upcoming multinational partnerships such as the Giant Magellan Telescope and Cherenkov Telescope Array.

At both ends of the spectrum, one can also find the specific cases of private consortia and treaty-based intergovernmental organizations. At one end, telescope consortia, privately funded and operated by universities, foundations and other entities, represent the "continuity" in the astronomy Big Science landscape over the last hundred years. These make up the majority of fully closed access facilities, for example, the W M Keck Observatory in Hawaii, the Palomar Observatory in California, and the Large Binocular Telescope Observatory in Arizona. Any open time is either through small amounts of discretionary "director's time", or in reciprocal time exchange programs with other facilities. This model is continuing with the next generation of optical facilities such as the Thirty Meter Telescope, planned for construction in Hawaii. On the other end, treaty-based intergovernmental astronomical or space organizations such as the European Southern Observatory (ESO) and European Space Agency (ESA) have a tradition of open access policies on their flagship research infrastructures such as the Very Large Telescope and the XMM-Newton space telescope, although some limited time may be set aside for members of the consortium and instrument builders. It remains to be seen where the latest astronomical intergovernmental organization, the Square Kilometer Array (SKA), currently in the process of defining its access policies, will fall on this spectrum.

This brief survey suggests that facilities with open access policies are in the minority; however, we are uncertain whether this is a result of historical conditions, community values or a deliberate strategy to increase the usage of key scientific instruments (Mauduit 2017). Nevertheless, most astronomy RIs that are constructed and operated nationally or through multinational agreements primarily serve their own national community or the community defined by an agreed partnership. Given that the model of such multipartner, closed RIs is the growing modus operandi, the trend is revealing dynamics that run contrary to the modern needs, values and demographic evolution of the astronomy community, or even forward-looking and aspirational statements about global open access made recently by the Group of Senior Officials (GSO) on global RIs, established under the auspices of the G7 (GSO 2017). In a sense, astronomy appears to be returning to the past situation of "haves" versus "have nots" (McCray 2000), but this time at an international level.

4. UNPACKING CONCEPTS UNDERLYING ACCESS AND RETURN ON INVESTMENT

A large number of actors are typically involved in the decision to construct a major observatory, from astronomers and science funders, to politicians, ministers and even heads of state. A range of scientific, industrial and political factors are considered, but particularly in the case where public funding is involved, project champions inevitably justify investments in terms of scientific and industrial return, amongst others, to the funding nation or agency. The concept of "return on investment" is commonplace in the policy language of research infrastructures and public funding in general. This reflects the broad influence of economic concepts on political thinking in the past century, particularly in post-World War II Western societies (Smith 2006). Yet what does "return on investment" mean in the context of access to research infrastructures and the endeavor of science, and to whom do the investment returns accrue?

There is a large extant literature on the benefits of investment in science and research infrastructures and how to evaluate those benefits (Georghiou 2015; Giffoni et al 2018; Guthrie et al 2013), which typically use cost-benefit analyses centered in economic theory to quantitatively calculate investment return. Many of the benefits of science facilities are not unique and are shared with other types of infrastructure development. Focusing on the scientific aspects of this sociological system, the concept of "return on investment" implicitly assumes that some or all parts of the system of science can be treated as "goods" in an economic system. Yet the scientific production system is greatly different from conventional market commodities and services. Leaving aside the direct external economic impacts from industry contracts and infrastructure development (Atkinson et al 2017; Florio et al 2016), the narratives on return on investment tend to emphasize two aspects with respect to the scientific system. First, return on investment is conceived in the form of national or institutional prestige generated by scientists affiliated with the funder gaining competitively awarded access, proprietary data, publications and credibility. Second, return is explained as a broad, indirect type of return to the public in the form of general scientific knowledge. In terms of access to a facility, the former obviously favors a closed model of access linked to contributions, whereas the latter, while not ruling out a closed model, can also favor an open system where access is decided by scientific excellence. In the following section, these two perspectives are explored in terms of their manifestations in the field of astronomy.

4.1 Access as a Private Good

The first type of return on investment is intrinsic to the hypercompetitive modern system of science. Observatory access, which nowadays is as prominent and important as grant awards or publications, can be interpreted as an "individual" or "private good". Such goods can only be consumed and used up by one consumer (rivalrous), and consumers cannot freely access the good (excludability). These features imply a natural sense of fairness in the producers of the good offering it in proportion to a contribution of a valuable resource such as money, technical skill or physical in-kind instruments or kit (Savas 2000). If the mass public funding of research and the general university system were different, one might imagine a system of private observatories that essentially competitively sold access time. In modern times, however, the high-technology requirements and large-scale nature of observatories coupled with the limited market and ability of university-employed researchers to pay directly for access essentially prohibit a market system, unlike the industrial users of various synchrotron facilities, for example (Hallonsten 2016). The "distribution" system of the access good comes with a twist that departs from the conventional economic approach. Rather than pure ability to pay, even in the private observatory systems access is often controlled by scientific review boards that evaluate the scientific merit of a proposal. In essence, the "payment" is made in advance by the funding facility, but access is determined by a principal investigator's scientific credibility. This individualized value is built on scientific reputation and record, but is also derived from the existence of "externalities" or goods in the broader scientific system such as openly available data and publications, a vibrant research community and public support of university and knowledge systems.

An individual's scientific value is drawn substantially but also indirectly from the availability of open data, which supports follow-on scientific findings, verification and confirmation of findings. In recent years, almost half of all papers published from data coming from Hubble have been from recycling archival data rather than new observations (White et al 2009). Apart from a restricted "proprietary" period of up to two years, which allows principal investigators to transfer their individually awarded access time into a first publication, the majority of facilities make their data archives public. Indeed, astronomy has led the world in open data, creating the virtual observatory system of common standards and formats that allows data from multiple different observatories to be combined and shared. Yet why should observatories fund this activity? This speaks to the egalitarian nature of astronomy described earlier in the chapter, where individual scientists recognize that their individual success and competitively awarded observatory access time depend on a broader system. Whether they stem from new observations or recycled data in archives, the generation of publications is an important return for an RI (Hallonsten 2013). Users of archive data are typically required to cite the archive, and even the instrument or telescope from which the data were collected, thus allowing facilities to track their productivity, potentially a way to justify, ex-post, the significant investment in building the facility in the first place.

The key good in the astronomy system and the primary vehicle of securing a return on funders' investments is the commodity of first-author, first publication of a novel scientific result. Yet, as a natural science dependent mainly on observations of events beyond the investigator's control, anyone with the right facilities and skills can conduct astronomy. This fact naturally favors a closed system of access to facilities and data to prevent "free-riders". The community view of free-riding in astronomy is where one astronomer uses data from another astronomer's competitively awarded access time and publishes a novel scientific result first. To prevent this, in addition to proprietary data, a system of negotiated access and collaboration has emerged, where large collaborations group around a certain instrument or an observation campaign and secure privileged access rights, data usages and first publication (see e.g. Gilmozzi et al 2016; Marchis et al 2016). A large body of political science work has shown that negotiated agreements within a system of defined rules is the optimal way to prevent free-riding and regulate access to common-pool goods (Ostrom 2005). This doesn't prevent rival teams from doing the same, however, and competition is fierce between groups. Nevertheless, is this "tribal" system of negotiated agreements to allocate a share of a highly rivalrous novel scientific finding an optimal system? For the case of particle physics, where a novel scientific result takes many years and involves substantial capital costs, complex machines and vast collaborations, the negotiated agreements now involve thousands of people (Castelvecchi 2015).

The scientific system, especially in astronomy, prioritizes credibility of individual scientists, research groups or instrument teams, by linking privileged access to facilities to publication, which in turn generates benefits for institutions and funders in terms of calculating research productivity, or bringing institutional prestige. This chain of investment benefits begins with privileged access to a facility, yet this access is controlled by gatekeepers who employ a range of rationales to justify selection: whether the user belongs to one of the negotiated agreements that exist to build an instrument or telescope; whether the user is affiliated with a funder; and finally the scientific excellence of a proposal. This dynamic fits well with the "old" Big Science in astronomy, where national funding agencies in small nations prioritize scientists of that nation, and where international groups collaborate to build and control access to a facility. As described in the previous section, an aspect of transformed Big Science in astronomy, however, has broken this dynamic in the form of the US OSP that has led to most federally funded ground and space facilities to be open, and also in Europe, where ESO and ESA maintain a tradition of generally open access. Yet the reasons for why the publicly funded South African Large Telescope and Gran Telescopio Canarias remain closed to restricted groups, whereas the publicly funded facilities in the National Optical Astronomy Observatory and ESO are open, remain unclear. A recognized deficiency in the economic theory of science is the lack of understanding of institutional mechanisms linking the private good of elite access to RIs (which bestows reputational and ultimately financial rewards on individuals and institutions), with broader externalities of the science system and the incentives and inducements present in the academic and funding institutions (Kealey and Ricketts 2014).

4.2 Access as a Public Good

The second concept of a return on investment is much broader and related to the idea of science as a public good (Moriaty 2011). Access to observatory facilities that allow the obtaining of scientific data and production of scientifically competitive and novel results eventually contribute to a further type of good: general scientific knowledge. The scientific system incentivizes scientists to publish their results, both to receive critical review and also to receive credit. In the competitive system of science, the publication of papers, attendance at conferences, service as graduate supervisors and other community activities are ways that individuals and institutions benefit from the original scientific findings (Latour and Woolgar 1986/79). The more novel findings attributed to a scientist or institution, the more credibility gained, which contributes to a higher likelihood of competitively winning observatory access. Yet, these individualized benefits also feed into the broader scientific landscape creating general scientific knowledge – a non-rivalrous and non-excludable collective good. Any number of users can access the knowledge at any time, via libraries, conferences or science outreach for example, while despite publication paywalls and conference costs, the university system generally supports wide distribution of scientific findings. In conventional economic theory, this would be considered as an externality to the individual transaction of funding an observatory and getting first access to data and publications. In the scientific system, however, this "externality" is fundamental to the whole endeavor.

While scholars have doubted whether scientific knowledge is truly a collective good due to its dependence on the level of contribution and ability to use knowledge (Kealey and Ricketts 2014), its collective nature is invoked to justify public funding for science. For those observatories that are publicly funded, this fact has inevitably seeped into the rationales and formulations of access policies. Many publicly funded telescopes are at least fully open to nationals of the funding country. But defining a return on investment for open access, even within a closed national system, is more challenging than the story for individualized return on investment. While in theory, facilities are ultimately judged on the quality of science that they produce and return to public knowledge, in reality, policymakers and funders do raise questions if substantial fractions of primary results are produced by scientists unaffiliated with the funding nation or agency. Even given the moral ideals of the discipline, astronomers have been known to point out the inherent unfairness of unaffiliated scientists reaping the scientific benefits without contribution (McCray 2000). Yet, exceptions can be found to this view. In the case of the Indian Giant Metrewave Radio Telescope, the open access policy was influenced by the need to stimulate local capability in astronomy. On the other hand, Chile inverted this approach by offering very favorable hosting conditions for observatories, in return for guaranteed time. Despite the two different approaches, access policies were driven by local development needs in terms of technical capability but also national political prestige, in addition to the localized benefits to national scientists.

This discussion aims to point out that access is not the only good in the scientific system of an observatory and instead is one of many goods, all of which are connected with the individualist return on investment aspect that leads to closed access policies. There are many examples of these goods other than access. Astronomy RIs invariably produce data, which are eventually made available for others to use. Service in the various governing committees and time-allocation panels are viewed as highly prestigious tasks, especially as they play a large part as gatekeeper to what the community views as scientific excellence. All of the large observatories employ astronomers in the tens or even hundreds in the case of ESO, ESA and the US National Optical Astronomy Observatory, to operate and manage the facilities, which confers inside knowledge, key skills, privileged access and even more stable careers than in academia. Many observatories run training programs for junior astronomers and offer their locations as hosts for fellowships or studentships in partnership with local universities. All of these aspects provide essential elements in the credibility cycle, yet most are not sought after in the same way as access to the facility. Furthermore, there are many second-order externalities of observatory knowledge products, including general scientific knowledge, human capital accumulation, technology spillovers, cultural benefits and science outreach. These are disconnected, however, from the access-credibility cycle relationship.

5. CONCLUDING DISCUSSION AND POLICY IMPLICATIONS

This illustration of access to RIs in the field of astronomy in the previous sections is indicative of broader trends in the landscape of Big Science. Policymakers face a variety of contradictions as facilities become larger and more costly, accumulate more partners across national borders, face requirements generally to increase global access and demands to show return on investment while producing the best science in combination with broader socio-economic impacts. In a broad sense, the access debate centers around a fundamental question about how to value investments made in RIs. Typically, policymakers conceptualize this value in terms of ensuring "fairness" for scientists affiliated with the funder, in terms of maximizing short-term scientific gain through publishing novel science results first, or in building their own scientific and technological capacity. The question remains, however, whether this approach is an optimal strategy for maximizing long-term scientific gain. Indeed, comparative analyses of synchrotron radiation user facilities show that publication metrics are poor indicators to value investments in complex socio-technical systems such as Big Science RIs (Hallonsten 2014a). These challenges demand a conceptual framework for understanding and formulating access policies, and to move beyond the quid pro quo view of access in return for contribution.

The future developments in the field of astronomy mentioned in this chapter illustrate a situation where major facilities will likely become increasingly closed, raising questions about the long-term sustainability of the science and the strategic implications for funding and participation in major projects. Similar issues are emerging in other fields that rely on major capital-intensive facilities. The issues surrounding the European X-Ray Free-Electron Laser

Facility in Hamburg, for example, have brought similar issues to bear about the nature of contribution versus access. The facility risks a situation where scientific use is disproportionate to the national shares set through higher-level political processes, which saw Germany and Russia take shares in the construction costs that dwarf all other contributions (Cramer, ch 3 in this volume). Despite a rebalancing applied by a scientific fair return policy that set the overall use relative to contribution, for smaller countries there is then a "perverted" incentive to seek the lowest contribution possible (Hallonsten 2012). These issues have been avoided in ESO, the European Organization for Nuclear Research (CERN) and ESA, which adopt no specific access policy relative to the contributions pre-defined by the national gross domestic product relative to other members (Cogen 2012). Yet in the case of ESO, the high cost of the Extremely Large Telescope is now raising questions about the formerly generous global open access policies (ESO 2019).

The short survey conducted in this chapter illustrates that, at least in the case of astronomy, there is no generalized set of principles that can be used to determine an access policy to a facility. Only the US OSP gets close to formulating some basic rationales and principles. The future of Big Science facilities under construction with restricted access such as the Cherenkov Telescope Array and SKA, however, is causing policymakers to reason more explicitly about OSP. The simultaneous construction of three 30 m class optical ground-based telescopes (ESO's Extremely Large Telescope and two US-led public-private consortia telescopes, the Giant Magellan Telescope and the Thirty Meter Telescope) and their remaining funding shortfalls, particularly for the US projects, has led policymakers to employ a closed access as an incentive tool to gain new partners (ESO 2019). With a view to both meeting the return on investment demand from domestic funding agencies and politicians, and also projecting the likely future landscape of highly competitive international facilities without any US involvement (e.g. the Extremely Large Telescope, Cherenkov Telescope Array and SKA), senior figures in the US community have suggested potential modifications to the OSP. Some have reimagined the OSP as a negotiated agreement with other international facilities that provides a common pool of open facilities to those affiliated with the nations and institutes in the agreement (private communication with the authors).

The recent IAU strategic plan foreshadows the requirement for global coordination and discussion:

> International efforts play a key role in driving all areas of astrophysics, and involve access to data and facilities as well as joint partnerships on large-scale instruments, observatories, and missions. Global cooperation and collaboration are increasingly important as the costs and preparation to build forefront facilities and missions escalate. International strategic planning is essential to explore how partnerships

> can be built and joint projects developed that could otherwise not be afforded, and to maximise the scientific return from these efforts. (IAU 2018: 19)

This could be envisaged as a high-level discussion between all top-facilities directors under the auspices of a neutral and global body such as the IAU, with the aim to produce a set of guidelines. In practice, creating a scientifically optimal and fair set of access policies, however, will likely require open international consultation and discussion within scientific communities.

The IAU statements are echoed in other international policy arenas. Despite the contemporary focus on impact and excellence, a normative theme is also present, which emphasizes the importance of international researcher mobility and broad access to international facilities. At the G7 summit in Okinawa in 2000, G7 science ministers released a joint statement that established the GSO, with a mandate to promote international cooperation on a variety of policy areas concerning research infrastructures, in particular the promotion of global transnational access (GSO 2017). The GSO has developed a self-assessment framework criteria for research infrastructures which, amongst other aspects, calls for research infrastructure policies to "reflect the global-Excellence-driven Access (gEA) paradigm through publication of a clear and transparent access goal. The goal should incorporate a peer-reviewed process that recommends access based on the most promising emergent ideas, regardless of the country of origin or the ability of the proposer to contribute financially" (GSO 2017: 15). These policy areas are also a focus of the Organisation for Economic Co-operation and Development's (OECD) Global Science Forum, and the European Commission, which shepherded the development of the European Charter on Access to Research Infrastructures (EU COM 2016). This policy intent is backed by serious funding: one of the stated priorities of the Commission's €80 billion Horizon 2020 framework program under the heading of "Open Science" is to support transnational access to research infrastructures and many funding calls have incorporated these goals.

Access policies in Big Science have emerged from ad hoc, bespoke local political and historical contexts, rather than from reasoned, principled analysis, and in spite of several high-level policies from the OECD, GSO, European Union and IAU that generally espouse the requirements for open access. Even Article 27 of the Universal Declaration of Human Rights of the United Nations General Assembly, which enshrines the right of everyone to "share in scientific advancement and its benefits" (UN 1948), seems to support an open access approach. Large astronomical research infrastructures could be considered as scientific advancements procuring the benefits of opening new windows into our Universe. While this may be a stretch, discussions to define the human right to science are ongoing and researchers at the forefront of the field have remarked that "scientists must have access to the materials necessary to conduct their research, and access to the global scientific community" (Wyndham and Vitullo 2018). Further scholarly research on this subject is critically needed to further understand the overall policy tensions and conceptual arguments that underpin the access question.

ACKNOWLEDGMENTS

Jean-Christophe Mauduit would like to acknowledge Dr. Niruj Mohan Ramanujam for his insightful comments and discussions on the topic.

REFERENCES

Adams J (2013) The fourth age of research. *Nature* **497** (7451): 557–60.

Atkinson D, R Wolpe and H Kotze (2017) *Socio-economic assessment of SKA phase 1 in South Africa, prepared as part of the Strategic Environmental Assessment (SEA) for the Phase 1 of the Square Kilometre Array (SKA) radio telescope, South Africa*. Available at www.skaphase1.csir.co.za/wp-content/uploads/2017/01/SocioEconomic-Assessment.pdf (last accessed December 17, 2019).

Castelvecchi D (2015) Physics paper sets record with more than 5,000 authors. *Nature News*, May 15.

Coccia M and L Wang (2016) Evolution and convergence of the patterns of international scientific collaboration. *Proceedings of the National Academy of Sciences* **113** (8): 2057–61.

Cogen M (2012) Membership, associate membership and pre-accession arrangements of CERN, ESO, ESA, and EUMETSAT. *International Organizations Law Review* **9** (1): 145–79.

Cruz-Castro L, K Jonkers and L Sanz-Menéndez (2015) The internationalisation of research institutes. In A Wedlin and M Nedeva (eds) *Towards European science: Dynamics and policy of an evolving European research space*. Edward Elgar Publishing, pp 175–98.

ESO (2019) *Report from Scientific Technical Committee, 93rd meeting, Garching, 2 and 3 May 2019*. Available at www.eso.org/public/about-eso/committees/cou/cou-150th/external/STC_632_recommendations_93rd_STC_meeting_May_2019.pdf (last accessed December 17, 2019).

EU COM (2016) European Charter for Access to Research Infrastructures. Available at https://ec.europa.eu/research/infrastructures/pdf/2016_charterforaccessto-ris.pdf (last accessed December 17, 2019).

EU COMPET (2018) *Council conclusions on "Accelerating knowledge circulation in the EU"*. Available at http://data.consilium.europa.eu/doc/document/ST-9507-2018-INIT/en/pdf (last accessed December 17, 2019).

Finkbeiner A (2018) The new era of multimessenger astronomy. *Scientific American* **318** (5): 36–41.

Florio M, S Forte and E Sirtori (2016) Forecasting the socio-economic impact of the Large Hadron Collider: A cost–benefit analysis to 2025 and beyond. *Technological Forecasting and Social Change* **112**: 38–53.

Georghiou L (2015) *Value of research.* Policy paper by the Research, Innovation, and Science Policy Experts, European Commission. Available at https://ec.europa.eu/research/innovation-union/pdf/expert-groups/rise/georghiou-value_research.pdf (last accessed November 15, 2019).

Giffoni F, T Schubert, H Kroll, A Zenker, E Griniece, O Gulyas, J Angelis et al (2018) *State of play: Literature review of methods to assess socio-economic impact of research infrastructures.* RIPaths project. H2020 grant 777563. Available at https://ri-paths.eu/wp-content/uploads/2018/03/D3.2_Report-on-stocktaking-results-and-initial-IA-model.pdf (last accessed December 17, 2019).

Gilmozzi R, L Pasquini and A Russell (2016) VLT/VLTI second-generation instrumentation: Lessons learned. *The Messenger* **166**: 29–35.

Grant B (2017) Philanthropic funding makes waves in basic science. Available at www.the-scientist.com/careers/philanthropic-funding-makes-waves-in-basic-science-30184 (last accessed December 17, 2019).

Griniece E, A Reid and J Angelis (2015) Evaluating and monitoring the socio-economic impact of investment in research infrastructures. Technopolis Group. Available at www.technopolis-group.com/wp-content/uploads/2015/04/2015_Technopolis_Group_guide_to_impact_assessment_of_research_infrastructures.pdf (last accessed January 13, 2020).

GSO (2017) *Progress report.* Available at https://ec.europa.eu/info/sites/info/files/research_and_innovation/gso_progress_report_2017.pdf (last accessed December 17, 2019).

Guthrie S, W Wamae, S Diepeveen, S Wooding and J Grant (2013) *Measuring research: A guide to research evaluation frameworks and tools.* RAND Europe.

Haas P (1992) Introduction: Epistemic communities and international policy coordination. *International Organization* **46** (1): 1–35.

Hallonsten O (2012) Continuity and change in the politics of European scientific collaboration. *Journal of Contemporary European Research* **8** (3): 300–19.

Hallonsten O (2013) Introducing "facilitymetrics": A first review and analysis of commonly used measures of scientific leadership among synchrotron radiation facilities worldwide. *Scientometrics* **96** (2): 497–513.

Hallonsten O (2014a) How expensive is Big Science? Consequences of using simple publication counts in performance assessment of large scientific facilities. *Scientometrics* **100** (2): 483–96.

Hallonsten O (2014b) The politics of European collaboration in big science. In M Mayer, M Carpes and R Knoblich (eds) *The global politics of science and technology*, Vol. 2. Springer, pp 31–46.

Hallonsten O (2016) Use and productivity of contemporary, multidisciplinary Big Science. *Research Evaluation* **25** (4): 486–95.

IAU (2012) *Astronomy for development: Building from the IYA2009 – strategic plan 2010–2020 with 2012 update on implementation.* Available at www.iau.org/static/education/strategicplan_2010-2020.pdf (last accessed December 17, 2019).

IAU (2018) *IAU strategic plan 2020–2030, 18 April 2018.* Available at www.iau.org/static/education/strategicplan-2020-2030.pdf (last accessed December 17, 2019).

Kealey T and M Ricketts (2014) Modelling science as a contribution good. *Research Policy* **43** (6): 1014–24.

Latour B and S Woolgar (1986/79) *Laboratory life: The construction of scientific facts*, 2nd ed. Princeton University Press.

Marchis F, P Kalas, M Perrin, Q Konopacky, D Savransky, B Macintosh et al (2016) Large collaboration in observational astronomy: The Gemini Planet Imager exo-

planet survey case. In A Peck, R L Seaman and C R Benn (eds) *Observatory operations: Strategies, processes, and systems VI*, Vol. 9910. International Society for Optics and Photonics.
Mauduit J C (2017) Open skies policies in astronomy: The growing need for diplomacy on the final frontier. *Science Diplomacy* **6** (2). Available at http://sciencediplomacy.org/article/2017/open-skies-policies-in-astronomy (last accessed December 17, 2019).
McCray W (2000) Large telescopes and the moral economy of recent astronomy. *Social Studies of Science* **30** (5): 685–711.
Moriarty P (2011) Science as a public good. In J Holmwood (ed) *A manifesto for the public university*. Bloomsbury Academic, pp 56–73.
National Science Foundation (NSF) (2014) *Report of the Astronomy and Astrophysics Advisory Committee*. Available at www.nsf.gov/mps/ast/aaac/reports/annual/aaac_2014_report.pdf (last accessed December 17, 2019).
National Science Foundation (NSF) (2018) *Science and engineering indicators 2018*. National Center for Science and Engineering Statistics, National Patterns of R&D Resources. Available at www.nsf.gov/statistics/2018/nsb20181/assets/1038/figures/fig04-04.pdf (last accessed December 17, 2019).
Ostrom E (2005) *Understanding institutional diversity*. Princeton Press.
Pandor N (2012) South African science diplomacy. *Science and Diplomacy* **1** (1). Available at www.sciencediplomacy.org/perspective/2012/south-african-science-diplomacy (last accessed December 17, 2019).
Pozza M (2015) Diplomacy for science: The SKA project. In L S Davis and R G Patman (eds) *Science diplomacy: New day or false dawn?* World Scientific Publishing, pp 87–106.
Savas E (2000) *Privatization and public partnerships*. Chatham House.
Smith K (2006) Economic techniques. In R Goodin, M Moran and M Rein (eds) *The Oxford handbook of public policy*, Vol. 3. Oxford University Press, pp 729–45.
Thomasson A and C Carlile (2017) Science facilities and stakeholder management: How a pan-European research facility ended up in a small Swedish university town. *Physica Scripta* **92** (6), 062501.
UN (1948) *Universal Declaration of Human Rights*. Available at www.ohchr.org/EN/UDHR/Documents/UDHR_Translations/eng.pdf (last accessed January 13, 2020).
Wagner C, T Whetsell and L Leydesdorff (2016) Growth of international collaboration in science: Revisiting six specialties. *Scientometrics* **110** (3): 1633–52.
White L, A Accomazzi, G Berriman, G Fabbiano, B Madore, J Mazzarella et al (2009) *The high impact of astronomical data archives*. Astro2010: The Astronomy and Astrophysics Decadal Survey, Position Papers, no. 64. Available at https://ui.adsabs.harvard.edu/abs/2009astro2010P..64W/abstract (last accessed December 17, 2019).
Wyndham, J and M Vitullo (2018) Define the human right to science. *Science* **362** (6418): 975.
Ziman J (1994) *Prometheus bound: Science in a dynamic steady state*. Cambridge University Press.

10. Is there an "iron law" of Big Science?

Olof Hallonsten

1. INTRODUCTION

Big Science is no longer only the domain of centrally planned scientific projects operated in an industrialized manner, but also increasingly includes facilities and resources used by ordinary-sized research groups (see Cramer et al, ch 1 in this volume). The investments necessary to establish these large research infrastructures are still big and politically conspicuous. Economic geographer Bent Flyvbjerg, who has studied the politics and sociology of "megaprojects", meaning infrastructure projects with a total investment cost of over $100 million (or €85 million), has identified an "iron law" of megaproject management, namely "over budget, over time, under benefits, over and over again" (Flyvbjerg 2017: 9). None of Flyvbjerg's writings, or the compiled books he has been the author of, pay any specific attention to megaprojects in science, a category that has been subject to some analysis by other authors (Eggleton 2017; Robinson 2019). The topic of this book begs the question of how well, if at all, the "iron law" applies to Big Science and the varieties of Research Infrastructures (RIs) in Europe that require large investments (e.g. more than €85 million), that are technically complex and that require planning, preparation and construction efforts stretching out several years in time.

This chapter therefore tries the notion of the "iron law" on a subset of RI projects that are listed in the roadmaps of the European Strategy Forum on Research Infrastructures (ESFRI) and that have an estimated investment cost exceeding €85 million and thus nominally, in Flyvbjerg's (2011, 2017) definition, are "megaprojects". A selection of 17 projects from the ESFRI roadmaps was made in order to get longitudinal data on cost estimates and time schedules, information that can be used to assess whether these projects have been completed "over budget" and/or "over time", and if there is a consistent pattern in the sample that warrants the conclusion that this happens "over and over again". The "under benefits" criterion of the "iron law" is left out of the analysis of this chapter due to the difficulty of assessing the benefits of RIs for science and society. The analysis is complemented by a detailed study, to give some more depth to the discussion, on one of the 17 projects in the sample, namely the neutron facility the European Spallation Source (ESS) under construction in Lund in southern Sweden. This analysis is based on the comprehensive history of the project in another publication (Hallonsten 2020b).

The purpose of this chapter is not to establish unequivocally whether or not Big Science and RIs that are of a certain size and complexity suffer from an "iron law". The core aim of this chapter is thus instead to discuss, based on the findings of the empirical analysis, whether, how and to what extent the "iron law" applies and what it means in a broader perspective, for the macrosociological study of the planning, funding, governance and organization of big science and RI projects in Europe. For reasons of simplicity, RIs that cost more than €85 million and thus qualify as "megaprojects" in Flyvbjerg's definition are identified as Big Science in this context. Although the empirical material was carefully chosen and is fairly representative (see a later section), it has limitations in terms of generalizability of the findings.

The chapter begins with a brief summary of the concept of the "iron law" of megaprojects, and its causes and dynamics, as theorized by Bent Flyvbjerg and colleagues. Thereafter, the selection of 17 projects from the roadmaps is outlined and discussed in terms of whether they have exceeded budgets and time schedules. In the sections to follow, the case study of the ESS project is presented and analyzed. The chapter ends with a concluding discussion and some policy implications.

2. THE "IRON LAW"

Through a series of publications, economic geographer Bent Flyvbjerg has established himself as a leading analyst of so-called infrastructure "megaprojects", which is defined as projects with a total investment cost of over $100 million, or €85 million, and which typically include tunnels, bridges, railways, highways and similar. The "iron law" is defined by Flyvbjerg and colleagues as: "over budget, over time, under benefits, over and over again" (Flyvbjerg 2017: 9–12), meaning that projects of this size regularly and with very few exceptions greatly exceed budgets and timetables. The claim is backed by a large number of examples.

As part of the analysis and synthesis of these findings, Flyvbjerg has also named a few explicit reasons for cost and timetable overruns that, hence, can be said to be the causal explanations of the law: (1) inherent risk and expectable uncertainty of long planning horizons; (2) complexity of decision making and planning processes, including goal conflicts and handling of conflicts of interest; (3) the use of novel, non-standard and thus untested techniques and designs; (4) overcommitment to specific design concepts on early stages that may create lock-ins that weaken alternatives that may be viable and cost-efficient; (5) enhanced risk of principal–agent problems due to the large sums of money involved; (6) significant changes of project scope or levels of ambition as time passes; (7) a tendency to avoid planning for unplanned events; (8) a risk that misinformation spreads about costs, schedules, benefits and risks; and (9) feedback loops where cost overruns as well as changed or lowered ambitions produce uncertainty over project viability and the eventual benefit of completion (Flyvbjerg 2011: 322).

Some minor adaptation of the "iron law" is necessary to make it work in the present context. Flyvbjerg's definition of the law is "over budget, over time, under benefits, over and over again". As noted in the introduction to this chapter, the "under benefits" criterion is omitted here, simply because any evaluation of the benefits of Big Science or RIs must be made *ex post* and preferably after several decades of scientific operation, and even then, the issue of "benefit" or "impact" is complicated (cf. Hallonsten 2016: 185ff; Hallonsten and Christensson 2017), perhaps particularly so in comparison with the infrastructure projects that the original "iron law" was developed to pertain to – bridges, tunnels, highways, railways and so on. What remains of the "iron law" to be analyzed here are thus "over budget", "over time" and "over and over again". The latter is a mere generalization of the other two; implying that budget and timetable overruns are common.

One important aspect of Flyvbjerg's identification of the "iron law" is that cost and timetable overruns are, to some extent, *relative*: What looks like real cost increases and timetable overruns due to any of the nine types of reasons listed above can also be the result of adjustment to reality, if early planning phases are characterized by overoptimism in cost and timetable estimations and reluctance to properly plan for unforeseen events. There are both political and technical reasons for such mis- or underestimation of costs and delays. Points (3), (6) and (7) above – untried technology, change of scope and levels of ambition, and avoidance to plan for contingencies – most evidently pertain to this. The fact is that it is close to impossible to foresee and make realistic estimations of costs and time schedules for all processes involved in these highly complex projects. Moreover, it is probably politically impossible to gain support for excessive investments simply on the basis of a very high contingency budget. Any proponent of an expensive project will probably lapse into underestimations of cost, either believing that they are correct and realistic or deliberately misleading decision makers and the public in order to increase the chances of getting their project funded, postponing the issue of dealing with cost increases and delays to other decision makers that will bear the responsibility later on. These successors at the helm, however, are not necessarily either able or willing to answer for the overoptimistic calculations and estimations made before their terms of office. It is therefore reasonable to expect that the logic of political decision making is counterproductive to responsible planning and budgetary work. It is unrealistic to expect politicians and campaigners to make reasonable and well-founded cost estimations during early political campaign phases. Moreover, the situation today is such that few or none of the decision makers remain in office long enough to have to take political responsibility for cost increases and delays, either those that are real or caused by too optimistic predictions in the initial decision-making phases.

3. THE 17 RESEARCH INFRASTRUCTURES AND THE "IRON LAW"

The "iron law" is time-dependent, because it hypothesizes that budgets and time schedules for megaprojects rarely hold which can only be proven or disproven with significant hindsight. This means that in order to assess whether it applies on a specific project or set of projects, it is necessary to have longitudinal data and preferably as long time series as possible.

The ESFRI roadmap, issued regularly by the advisory body to the European Commission that is called ESFRI (Bolliger and Griffiths, ch 5 in this volume), contains lists and extensive descriptions of RIs "of pan-European interest and of scientific excellence" that have "reached different degrees of maturity" (ESFRI 2006: 22). Though it has been shown that the collection of RIs listed in one or several of the ESFRI roadmaps (five editions have been published to date) is varied to the degree that it can be questioned whether it comprises a relevant category other than in a policy sense (Hallonsten 2020a), the ESFRI roadmap is still the main policy document in Europe concerning RIs. The projects listed in a particular edition of the roadmap are thus reasonable to consider the most prioritized in Europe at the given point in time, and therefore a workable starting point for obtaining a sample of RIs to analyze in this chapter.

The first edition of the ESFRI roadmap, published in 2006, contained a total of 35 projects, of which 26 were estimated to have investment costs of €85 million or more, and thus qualify as "megaprojects" in Flyvbjerg's definition. Twelve years later, in 2018, the fifth edition of the ESFRI roadmap was published. In addition to its ordinary list of strategically important RIs, the report also contained a list of "landmarks", which means projects that have previously been listed in a roadmap edition but have now been taken into operation or reached "advanced stage of construction" (ESFRI 2018: 12). Of the 26 "megaprojects" in the 2006 roadmap, 17 were listed as "landmarks" in the 2018 report, and were therefore chosen as a sample of projects to be analyzed in this chapter.

The 17 projects display a significant variety in most available taxonomies. Six out of seven ESFRI categories are represented, with projects distributed as follows: Biomedical and Life Sciences (five projects); Astronomy, Astrophysics, Nuclear and Particle Physics (three); Computer and Data

Treatment (one); Energy (one); Environmental Sciences (three); and Material Sciences (four). All except social sciences and humanities are included, which is hardly surprising given that RIs in this category hardly ever cost as much as €85 million.

Various taxonomies for RIs were proposed by Hallonsten (2020a), and they apply as follows on the present sample: Nine of the projects are "instruments", i.e. "technological systems or setups that are used in experimentation or measurement"; five are "observatories", i.e. "technologies for studying real-world phenomena" including "astronomical observatories but also e.g. seafloor or atmosphere observation and measurement technology"; and three are "repositories", i.e. "collections of material or data that can be used in research" including "databases, biobanks and the like". There are no "vessels" among the projects. Eight are single disciplinary and nine are multidisciplinary; nine are centralized and eight are distributed. Except in terms of costs, these 17 European RIs make up a rather neatly representative sample.

The 17 projects are listed in Tables 10.1 and 10.2 along with information from the 2006 and 2018 ESFRI roadmap reports concerning estimated costs and timetables.

Out of the 17 projects, ten have had their estimated price tags increase between 2006 and 2018. Some projects – JHR, ESS, XFEL and ELI – stand out with estimated cost increases of over half a billion. In relative terms, the cost increases of the ELI and JHR are most dramatic, with over 450 and 250 percent, respectively, and there are also two projects (EATRIS, SPIRAL2) that have had their estimated investment costs roughly doubled in the period. Interestingly, however, six projects actually lowered their cost estimates in the period, with some even cutting them by half or more (LIFEWATCH, ELIXIR, ICOS).

Table 10.2 shows that none of the 17 projects in the sample were completed at the time envisaged in the 2006 roadmap. While there seems to be some variation in the sample concerning cost increases, and several examples of lowered costs as projects mature, the pattern of delays is unequivocal, with INSTRUCT, SKA, JHR, ESS and SPIRAL2 standing out. Summing up this part of the empirical analysis, it can therefore be concluded that "over time" is ubiquitously present in the sample, whereas "over budget" seems less obvious across the board but conspicuous for some specific projects. "Over and over again" is significant at least for the delay variable, though not as evident for "over budget".

It should, however, be noted that the information in Tables 10.1 and 10.2 and the inferences made from them are based on data that is incomplete in (at least) three respects. First, estimations of construction costs and year of start of operation are taken from the 2006 ESFRI roadmap, which was compiled in committee work over two years (Hallonsten 2020a; Bolliger and Griffiths,

Table 10.1 Sample of 17 projects sorted on change in cost estimates between 2006 and 2018 (descending order)

Name	Estimated construction cost (million €) 2006	Capital value (million €) 2018	Change (million €)
Jules Horowitz Reactor (JHR)	500	1800	+1300
European Spallation Source (ESS)	1050	1843	+793
Extreme Light Infrastructure (ELI)	150	850	+700
European X-Ray Free-Electron Laser (XFEL)	986	1490	+504
Extremely Large Telescope (ELT)	850	1120	+270
European Infrastructure for Translational Medicine (EATRIS)	255	500	+245
Système de Production d'Ions Radioactifs en Ligne de 2e génération (SPIRAL2)	137	281	+144
Integrated Structural Biology Infrastructure (INSTRUCT)	300	400	+100
Institute Laue-Langevin (ILL) upgrade	160	188	+28
Biobanking and Biomolecular Resources Research Infrastructure (BBMRI)	170	195	+25
Partnership for Advanced Computing in Europe (PRACE)	200–400	250	
European Multidisciplinary Seafloor and Water Column Observatory (EMSO)	150	100	−50
Integrated Carbon Observation System (ICOS)	255	116	−139
The European Infrastructure for Phenotyping and Archiving of Model Mammalian Genomes (Infrafrontier)	320	180	−140
Square Kilometre Array (SKA)	1150	1000	−150
Science and Technology Infrastructure for Biodiversity Data and Observatories (LIFEWATCH)	370	150	−220
European Life-Science Infrastructure for Biological Information (ELIXIR)	550	125	−425

ch 5 in this volume), and therefore all the projects had been under planning for years before 2006. This means that cost increases and delays in the right-most columns of the two tables may be both under- and overestimated, since estimations both more and less accurate could have been made earlier that we have no information about. Second, similarly, six of the 17 projects are still not completed and their respective start of operation listed in Table 10.2 are still estimations, which means their delays can increase further, as can in principle also their investment costs. Third, the sample is limited and maybe also biased; there are of course RI projects in Europe and elsewhere that pass the €85 million "megaproject" threshold but are not in the sample, because ESFRI

Table 10.2 Sample of 17 projects sorted on apparent delay in years (descending order)

Name	Estimated start of operation in 2006 roadmap	Start of operation, actual (*) or estimated in 2018 roadmap	Change (years)
Integrated Structural Biology Infrastructure (INSTRUCT)	2007	2017*	10
Système de Production d'Ions Radioactifs en Ligne de 2e génération (SPIRAL2)	2011	2019	8
Jules Horowitz Reactor (JHR)	2014	2022	8
European Spallation Source (ESS)	2017	2025	8
European Life-Science Infrastructure for Biological Information (ELIXIR)	2007	2014*	7
Square Kilometre Array (SKA)	2014–20	2027	7–13
The European Infrastructure for Phenotyping and Archiving of Model Mammalian Genomes (Infrafrontier)	2007	2013*	6
Integrated Carbon Observation System (ICOS)	2010	2016*	6
Extremely Large Telescope (ELT)	2018	2024	6
Biobanking and Biomolecular Resources Research Infrastructure (BBMRI)	2009	2014*	5
European Multidisciplinary Seafloor and Water Column Observatory (EMSO)	2011	2016*	5
Extreme Light Infrastructure (ELI)	2013	2018*	5
European X-Ray Free-Electron Laser (XFEL)	2013	2017*	4
Institute Laue-Langevin (ILL) upgrade	2012–17	2020	3–8
Partnership for Advanced Computing in Europe (PRACE)	2008	2011*	3
European Infrastructure for Translational Medicine (EATRIS)	2010	2013*	3
Science and Technology Infrastructure for Biodiversity Data and Observatories (LIFEWATCH)	2014	2017*	3

did not include them in their 2006 roadmap or note them as landmarks in their 2018 roadmap. It is possible that presence on the ESFRI list pushes projects somewhat towards completion, which could presumably limit the delays somewhat in comparison to a broader and even more representative sample. Given that all projects included in the 2006 roadmap also got "preparatory phase" funding from the European Commission, this possibility is perhaps not only theoretical but real.

On the other hand, all samples are biased and as shown in previous publications and other chapters in this book (Hallonsten 2016, 2020a; Cramer et al, ch 1 in this volume; Moskovko, ch 6 in this volume), the categories RIs, Big Science and so on are ambiguous, amorphous and diverse. In order to balance this deficit in the current context, and enable inferences that can be of assistance in the qualitative discussion over the presence of an "iron law" in Big Science, the chapter will now turn to a case study to be analyzed in some more depth.

4. THE EUROPEAN SPALLATION SOURCE AND THE "IRON LAW"

The ESS is one of the 17 projects analyzed in the previous section. It is a European Big Science project that follows the tradition of intergovernmental political and scientific collaboration among universities and institutes across Europe to pool their resources and jointly build internationally competitive research facilities (Cramer, ch 3 in this volume). The ESS is currently run by a 15-country collaboration and co-hosted by Sweden and Denmark. It has been under planning since the early 1990s, when it was proposed as the next project to retain and enhance the internationally leading position of European neutron users. The ESS is a facility for neutron scattering, an experimental technique that allows scientists from a variety of fields (with some dominance of physics) to study the molecular structure of matter by using neutrons as a probe, much like the use of x-rays for similar purposes in synchrotron radiation and free electron laser facilities (for comprehensive descriptions, see Hallonsten 2016: 241ff). Its basic technological set-up includes a linear accelerator that powers a target station that produces high-intensity neutrons that are used in the experimental work.

The ESS was first proposed in 1990, and in 1993 a number of European institutes with strong capacity in neutron instrumentation and neutron use began joint work on a design proposal. The basic technical design and science case was presented in 1997, and at this time the facility was projected to cost €1500 million and be ready to start operations a decade later. After an endorsement of the project by the Organisation for Economic Co-operation and Development (OECD) in 1999, the loosely knit network of research organizations in Europe with an interest in neutrons hoped for a political agreement on the future location for the ESS, and its financial and organizational details. But in 2002/2003 the German and United Kingdom governments withdrew their support, and the project practically imploded politically. However, the withdrawal of the larger countries also opened a window of opportunity for some of the smaller and comparably unlikely candidates, including ESS-Scandinavia, a joint Danish–Swedish hosting bid which eventually found itself in competition with ESS-Bilbao, in Spain, and ESS-Hungary, which proposed a site in Debrecen in eastern Hungary. In May of 2009, after considerable time had passed and a gradually intensifying campaign involving the local and national governments that backed the three respective sites, a sufficient number of European countries aligned with the ESS-Scandinavia proposal and declared that Lund was to be the site for the future ESS. By then, technology had progressed significantly, and the ESS concept had also been slimmed down. Instead of the two so-called target stations that had been included in the original design (a long pulse and short pulse target), the ESS would only have one target station and thus be both less costly and scientifically more specialized and better adapted to the needs of the European scientific communities. In the years after the 2009 site decision, a comprehensive design update was undertaken, in parallel with negotiations between the prospective partner countries, who agreed on a funding solution in the summer of 2014. The new ESS design was published in 2013, and with it a new cost estimate of €1843 million (in 2013 prices). Construction started on site in September of 2014, and in October of 2015, legally binding agreements were made as part of the transition of the ESS organization from limited liability company to European Research Infrastructure Consortium (ERIC) (see Moskovko, ch 6 in this volume) with 11 members (Czech Republic, Denmark, Estonia, France, Germany, Hungary, Italy, Norway, Poland, Sweden, Switzerland), extended by the accession of the United Kingdom in 2016, and Spain in 2018. According to current estimations (2020), the facility will open to users in 2023, with an initial set of 16 instruments, which is six fewer than planned and detailed in the 2013 design (Hallonsten 2020b). As part of the planning and construction work, ambitions have also been lowered concerning the power of the accelerator, due to "ballooning costs", according to reports (Cartlidge 2017).

In Tables 10.1 and 10.2, the ESS is found among the projects with the ostensibly greatest cost increases and delays. But as the brief history in the previous paragraph shows, the data in the available ESFRI roadmaps is incomplete.

First, it should be noted that the estimated year of start of operation in the 2006 edition of the ESFRI roadmap was not the first such estimation made – as part of the 1997 ESS design the estimate was for the facility to be operational in 2007 – and so another ten years shall be added, and the total delay of the project, measured this way, is suddenly 18 years. When opening to scientific use in 2023, the ESS will have been under planning and construction for 33 years, which itself seems to be something of a record delay. Previous analyses have concluded that the delay is mainly due to inefficient or even inexistent policymaking frameworks and institutions in Europe especially compared to the United States, and so there is a specific political feature of Europe that aggravates the "over time" feature of the "iron law" at least for collaborative Big Science on this continent (Hallonsten 2014; Cramer 2017). Europe never had a structure in place that would enable the timely planning, decision making, funding and execution of Big Science like that within the system of United States National Laboratories or within individual European countries such as France, Germany and the United Kingdom. The documented delays of the ESS project were due mostly to this peculiar political situation in Europe; forced to collaborate to mobilize the critical mass in terms of competence to be able to compete on a global level, European countries do not have the mechanisms in play to enable such collaborations to be routinely and timely set up. While the European Union has made some efforts in the area (see Cramer, ch 3 in this volume; Moskovko, ch 6 in this volume; Bolliger and Griffiths, ch 5 in this volume; Ulnicane, ch 4 in this volume), it still plays a limited role.

Second, it should be noted that the cost estimate made in the 2006 edition of the ESFRI roadmap, of €1050 million, pertained to a slimmed-down version of the ESS, with only one target station and thus significantly smaller than the one outlined in 1997 at a cost of €1500 million. This adds – importantly – another dimension to the whole issue of cost increases or decreases, namely changes in ambition and scope. It is in principle possible that all the projects in Table 10.1 have had their scientific and technical ambitions lowered or heightened and that this is the major reason for the changes in their costs – all figures in column 2 of Table 10.1 are *estimations*, good or bad, which we will return to in the concluding discussion of this chapter. The increase of the costs of the ESS from €1050 million in 2006 to €1843 million in 2018 do, however, refer to a design of the ESS on the same or at least very similar level of ambition (Hallonsten 2020b).

In a first step, therefore, we can conclude that the ESS project is running "over time" and "over budget". But there is more to this story that can be of interest for the discussion.

A slight shift of perspective, to the Swedish view of the ESS, can be useful. When Lund was chosen as the site for the ESS at a meeting in May of 2009, this was made on the basis of a pledge by the governments of Sweden, Denmark and Norway to jointly cover 50 percent of the future construction costs of the facility. This pledge was made in *blanco*, meaning irrespective of future cost increases. As a comparison, the 2014 funding solution agreed by 13 partner countries and confirmed a year later in the legally binding ERIC statutes contained specific sums in million € in 2013 prices, which means that the agreement shielded the signatories from future cost increases and put the issue of covering for such increases up for negotiation. Of the 50 percent pledged in 2009 by Sweden, Denmark and Norway, 35 percent was to come from Sweden, which meant that the ERIC statutes of 2015 detail a Swedish financial contribution to the construction of the ESS of €645 million (European Spallation Source 2015/18). But this was far more than envisaged when the Swedish government announced its intention to get the ESS located to Lund, in

2007, and began the campaign on the European level. At that time, the promise was 30 percent of construction costs estimated at €1200 million, meaning a Swedish commitment of €360 million (Swedish Government 2007). The total estimated price tag was increased to €1478 million by the campaign in 2008, based on tentative design work, and the Swedish bid was consequently raised, but still some months after the 2009 site decision, when the Swedish government presented its budget bill to parliament and there laid out the plan for the financing of the Swedish share of the ESS facility; it still used the several-years-old estimate of a total ESS construction cost of "approximately 11 billion SEK" or €1200 million (Swedish Government 2009: 70–1). By then, the Swedish bid had been raised from 30 to 35 percent, and four years later, the total construction costs were set at €1843 million, but in 2013 prices, which means that with inflation (which was negligible in the Euro zone in 2013–15, but has oscillated between 1 and 2 percent thereafter), the sum is probably today (2020) around €2000 million. Moreover, real cost increases seem also to have occurred, most notably on the side of civil construction, and a cost increase of 980 million SEK (which corresponds to approximately €100 million) is apparently supposed to be borne by Sweden alone (Swedish Government 2016: 180), taking the total bill for the Swedish taxpayers to at least €750 million (€645 million as detailed in the ERIC statutes, plus inflation, plus €100 million extra). In other words, for Sweden alone, the estimated contribution to the construction of the ESS doubled between 2007 and today.

A previous analysis sought predominantly political explanations for this cost increase, claiming that Sweden was "unprepared" for this type of commitment, and showing how a complicated mix of direct funding and channeling of expenses through governmental agencies made tracing of the funding streams complicated (Hallonsten 2015). There are also testimonies of shady deals concerning contributions to the construction costs made between the government, the ESS-Scandinavia campaign, regional authorities and the university in Lund, which suggests that the government was unprepared or unwilling to fully take responsibility for its commitment to the ESS and what it could entail after cost increases and delays (Hallonsten 2020b). But this is of course not the sole or main reason for the cost increases. The design update, which enabled more realistic calculations of future costs, and the in *blanco* promises of percentages of the eventual construction budget, are also two other important factors.

5. DISCUSSION

It has already been concluded that in a general sense, the "iron law" applies on the 17 RIs analyzed, with some exceptions. A qualitative analysis of the logics and causalities of the processes involved also suggests that the "iron law" should apply. It is reasonable to expect several of the nine causal mechanisms of the law cited earlier in the chapter to be applicable on research infrastructure projects, which often are technically cutting-edge, have long planning horizons, are dynamic and changeable, and less well understood from policymakers' point of view. On the other hand, whether or not the learnings of the ESS case analyzed in some more depth in the previous section are generalizable and expected to be valid for the 17 projects from the ESFRI roadmaps or a broader set of RIs is doubtful. This is especially the case when considering the enormous heterogeneity of this group and how it seems to preclude any cross-case comparison without the a priori application of some differentiating definitions or categorizations (Hallonsten 2020a; see also Cramer et al, ch 1 in this volume). A safe conclusion concerning this specific issue is to suggest that case studies of similar depth and detail as the one accounted for here must be made also for other projects, in order to find out whether the "iron law" applies on them and thus on RIs of other kinds. The rather clear results of the analysis of the ESS case – that cost overruns amount to a doubling of the Swedish commitment, and that the inertia and lack of structure for the political decision-making processes in Europe delayed the ESS project by decades (see below) – are perhaps in themselves generalizable. Readers will have to judge for themselves, based on the following analytical conclusions drawn from the general patterns of the 17 projects, and the specifics of the ESS case.

In a first step, specifically for the ESS, two parts of the "iron law" undoubtedly hold, namely "over time" and "over budget". Considering specifically the "over time" part, it should be repeated that the key reason for the incredible delay of the whole project – it took almost a quarter of a century from idea and first proposal to start of construction – is most of all due to the politics of European collaboration in Big Science, as discussed briefly in this chapter and by Cramer (ch 3 in this volume). There is nothing to suggest that the ESS project, once the design work had started in 2009, or after the start of construction in 2014, suffered any unexpected delays, which might suggest that once European governments agreed on the funding, the threat of the "over time" feature of the "iron law" was neutralized, although deeper studies of the case are necessary to give a definite result on this account. The information available about the 17 projects analyzed in the first half of this chapter give no information on this point, but detailed case studies of these – certainly outside the scope of the current chapter but a promising focus for future research – can be the source of more generalizable findings.

Worse are, of course, the seeming threats of the "over budget" part. Here is a clear difference between the 17 projects and the ESS case. Among the former, at least six projects seem to have *lowered* their estimated construction costs in the time period covered. Whether or not lowered technical or scientific ambitions (similar to the downscaling of the ESS between 1997 and 2007, which took the price tag down from €1500 million to roughly €1000 million) are part of the explanation is not known but must of course be the subject of studies on a case-to-case basis.

The "over budget" part of the "iron law" in the ESS case is a better source of information to feed into an interesting discussion. Clearly, as seen in Table 10.1 and noted in the previous section, the ESS project got its estimated construction costs increased by €793 million in 2006–13, without any significant change in ambitions and scope of the project, but based on detailed design work. This brings to light that Big Science projects like the ESS are by nature technologically unpredictable, possibly more so than the generic infrastructure "megaprojects" analyzed by Flyvbjerg and colleagues. It is simply not possible to know, to a sufficient extent, the details of the technical installation that will be made as part of construction of the facility so that a reliable cost estimate can be made, before a detailed technical design has been developed. The 2006 cost estimate for the ESS, of €1050 million, was therefore by nature unreliable or even unrealistic, as was the 2007 estimate used by the Swedish government (of €1200 million). The detailed technical design for the ESS facility was not finalized until 2013, over three years after the site agreement of May 2009 and only after the ESS organization had grown to hundreds of employees – who else would undertake the design work? – and thus significant investments had already been made. This is probably true for all Big Science projects, and most RI projects, although it is reasonable to expect that the variations are at least as great as the variations in all other aspects (see Cramer et al, ch 1 in this volume, and Hallonsten 2020a). But the six projects in Table 10.1 that *lowered* their costs should probably be taken as an indication that unrealistic cost estimates can just as well be *overestimations*, although this needs to be analyzed in detailed case studies.

The political aspect is just as important, and the following discussion therefore borders on policy implications/recommendations. The lesson from the previous paragraph is that any cost estimation made at an early stage as part of the political process of anchoring a project and mobilizing support and line-item funding in the long-term financial planning of a government or similar, is by nature incorrect. Political logic also makes it more likely that cost estimations are unrealistically low, since it can be questioned whether policy- and decision makers, and the campaigners for projects that they unavoidably have to rely in part on, have any reason not to try to present costs smaller than perhaps would be motivated and reasonable if making responsible and realistic forecasts. Therefore, a warning or instruction should be issued to the policy- and decision makers involved in coming projects, that they include a greater contingency right from the start, and speak with greater honesty of the foreseen levels of costs.

The evidence from the current case as analyzed in the previous sections, and that can form the basis of such policy advice, is clear and convincing. But the recommendation appears to run counter to the logic of politics. First, it should be acknowledged that any politician at any time wants to present as favorable a picture of a coming investment and project as possible. Second, it is unrealistic to assume that any politician would have any success if presenting a large investment by referring to first a cost estimate and then to the fact that this cost estimate most likely is too low and needs to be complemented by a substantial contingency. This would be tantamount to asking for a large chunk of money in *blanco*. The issue is not to be resolved within the scope of this chapter. This can only be done through the identification and formulation of the problem on the basis of a relevant case and convincing data and analysis. Exceptions clearly also exist.

What we arrive at, as the conclusion of the chapter's analysis and discussion, is a purified or specialized version of the "iron law", possibly apt to name the "iron law of Big Science", and phrased as follows: Given the technological uncertainties involved in these types of projects, which cannot be avoided and which has very much to do with the long planning horizons, costs will unavoidably exceed early estimations and make it appear as if the projects go 'over budget". The special European situation deserves a separate analysis, and has also been subject to such (Cramer, ch 3 in this volume). But in a general sense, the problem is deepened by the political logic that also has to do with long planning horizons – policy- and decision makers involved in the early stages seldom or never have to stay on and answer for later cost increases – namely that projects are "sold" to governments, legislatures and taxpayers at too low levels. Ultimately, of course, this version of the iron law will lead to a credibility crisis for the Big Science facility projects themselves, unless the dynamics of the processes of their genesis and completion – political and technical – are better understood and better communicated.

REFERENCES

Cartlidge E (2017) Rising costs hamper mega-neutron beam facility. *Science* July 24.

Cramer K (2017) Lightening Europe: Establishing the European Synchrotron Radiation Facility (ESRF). *History and Technology* **33** (4): 396–427.

ESFRI (2006) *European roadmap for Research Infrastructures*. Report. Available at www.esfri.eu/sites/default/files/esfri_roadmap_2006_en.pdf (last accessed November 13, 2019).

ESFRI (2018) *Strategy report on Research Infrastructures*. Available at http://roadmap2018.esfri.eu (last accessed November 13, 2019).

Eggleton D (2017) *Examining the relationship between leadership and megascience projects*. Doctoral thesis, University of Sussex.

European Spallation Source (2015/18) *Statutes of the European Spallation Source ERIC*. Annex to Commission Implementing Decision (EU) 2015/1478 of 19 August 2015

on setting up the European Spallation Source as a European Research Infrastructure Consortium (European Spallation Source ERIC). Consolidated version, including amendments of 10 June 2016, 26 April 2018, 19 June 2018.
Flyvbjerg B (2011) Over budget, over time, over and over again: Managing major projects. In P Morris, J Pinto and J Söderlund (eds) *The Oxford handbook of project management*. Oxford University Press.
Flyvbjerg B (2017) Introduction: The iron law of megaproject management. In P Morris, J Pinto and J Söderlund (eds) *The Oxford handbook of project management*. Oxford University Press.
Hallonsten O (2014) The politics of European collaboration in big science. In M Mayer, M Carpes and R Knoblich (eds) *The global politics of science and technology*, Vol. 2. Springer.
Hallonsten O (2015) Unpreparedness and risk in big science policy: Sweden and the European Spallation Source. *Science and Public Policy* **42** (3): 415–26.
Hallonsten O (2016) *Big Science transformed: Science, politics and organization in Europe and the United States*. Palgrave Macmillan.
Hallonsten O (2020a) Research infrastructures in Europe: The hype and the field. *European Review* **28** (4): 617–35.
Hallonsten O (2020b) *The campaign: How a European Big Science facility ended up on the peripheral farmlands of southern Sweden*. Arkiv Academic Press.
Hallonsten O and O Christensson (2017) *An ex post impact study of MAX-lab*. Lund University.
Robinson M (2019) *Science mega-project communities; Mechanisms of effective global collaboration?* Doctoral thesis, Durham University.
Swedish Government (2007) *Regeringen vill bygga stor europeisk forskningsanläggning – partikelacceleratorn ESS blir störst i världen*. Press release from the Ministry of Education, Swedish government, February 26.
Swedish Government (2009) *Swedish governmental bill 2009/10:1, budget bill for 2010, expenditure area 16.*
Swedish Government (2016) *Swedish governmental bill 2017/18:1, budget bill for 2018, expenditure area 16.*

11. Keeping a Research Infrastructure alive: Material, social, and political work at the Institut Laue-Langevin

Beatrice D'Ippolito and Charles-Clemens Rüling

1. INTRODUCTION

Research Infrastructures (RIs) play an important role in European science policy and practice (Cramer et al, ch 1 in this volume; Ulnicane, ch 4 in this volume; Lozano et al 2014), yet relatively little is known about the social and political processes that drive their organizational and scientific development over time. Previous studies have focused on the creation and early development of RIs (Hallonsten 2015; Westfall 2008), or on the consequences of conducting research in the context of RIs for the internationalization and increase in interdisciplinary research (Elzinga 2012; Weinberg 1967). Some have also explored the financial difficulties these organizations may encounter in maintaining full functionality over time (Crease 2001, 2005). This chapter sheds light on the efforts addressed towards maintaining an RI in operation over an extended period of time; besides some exceptions discussing the extent to which the science becomes mature (Holmberg 2013; Westfall 2008) or the provision of decommissioning projects (Brumfiel 2004), how RI lifespans develop has not been fully addressed by previous research.

Even though RIs are typically established for a limited time, previous research argued that the organizations hosting an RI can be kept alive through the development of new missions and facilities (e.g. Hallonsten and Heinze 2012). In cases of *mission-driven* facilities, such as the European Organization for Nuclear Research (CERN), exceptional longevity has been achieved through the periodical establishment of entirely new research programs and associated instruments (i.e. via mission evolution), such as the Large Hadron Collider (LHC) or its projected successors, the International Linear Collider (ILC) or Compact Linear Collider (CLIC) (Banks 2014). In contrast, efforts to maintain *user-oriented* RIs such as in our case the Institut Laue-Langevin

(ILL), which have originally been established to provide wide ranges of external users with experimental resources of a well-defined nature (e.g. synchrotron radiation, neutron beams), typically focus on periodical technical upgrades. Examples include the work conducted on the European Synchrotron Radiation Facility upgrade program "Extremely Brilliant Source", or the evolution of the research program of the Stanford Linear Accelerator Center from particle physics to astroparticle physics, cosmology and multidisciplinary photon science, while maintaining its name and location (Heinze and Hallonsten 2017). Such renewal can ultimately go beyond technical upgrades to include complete retooling and orientation towards related disciplines (Westfall 2008).

An important justification for funding RIs is that they house expensive and powerful scientific instruments, generally considered too large to be built and managed by single university laboratories (Westfall 2010). However, whether these facilities warrant sustained investment is a question that is regularly debated by both scientists and policymakers (Crease 2001, 2005), and sustained funding might be related to RI flexibility and capacity to adapt to changing political and scientific demands. The operation of RIs over time thus depends not only on technical aspects related to the lifespan of an equipment, but also to a large extent on the political processes surrounding scientific endeavors (Holmberg 2013).

In this chapter, we focus on the efforts made by the ILL located in Grenoble, France, commonly considered the world's strongest and most versatile neutron source for science (Hallonsten 2016a; Rush 2015), to maintain its operations running for more than 50 years after its establishment in 1967 by France and West Germany. The ILL has been subject to multiple upgrades and improvement programs over the years (Arai and Crawford 2009). Nevertheless, the European Spallation Source (ESS) currently under construction in Lund, Sweden, can be considered the facility that might one day replace the ILL, questioning its continued operation. With the aim of contributing to the study of the development over time of RIs and institutional persistence, our chapter connects with existing literature that has examined the capability of RIs to adapt and renew their institutional and organizational remit (Hallonsten and Heinze 2012, 2015). In doing so, we analyze different domains of institutionalization, that is, physical infrastructures, scientific fields and science policy, and how they feed into each other.

2. SETTING THE SCENE: THE ILL AS A RESEARCH INFRASTRUCTURE

The use of neutron scattering for materials science developed after World War II across numerous countries with the construction, during the 1960s and

1970s, of neutron sources of various sizes. Among these research reactors, the ILL, which began its operation in the early 1970s, achieved the highest flux and rapidly became a reference point for neutron scattering research. The two main European neutron sources in operation, the ILL – considered "the world's flagship facility" – and ISIS (no acronym) located near Oxford in the United Kingdom (UK) which is the first pulsed neutron source, as compared to reactor-based sources, have been competing over the past decades on "instrument excellence, scientific output and service to users on both sources" (ESFRI 2016a: 5–6).

The ILL was established by an intergovernmental agreement between France and West Germany in 1967, with the UK joining in 1973. Since then, ten additional countries (Austria, Belgium, Denmark, Italy, Poland, Slovakia, Spain, Sweden, Switzerland and the Czech Republic) have joined the ILL as scientific members. The ILL provides expertise and facilities – over 40 scientific instruments for neutron scattering – to more than 1500 scientific users from over 40 countries annually, running an annual budget of about €100 million, and representing an overall investment estimated at €2 billion (ESFRI 2016a: 60).

Among scientists, the ILL has been recognized as the "best reactor source optimized for neutron scattering applications" on a worldwide scale (Arai and Crawford 2009: 14), and its organizational model has served as a blueprint for the design of other large RIs, such as the European Synchrotron Radiation Facility (ESRF) (Hallonsten 2009). As a result, the ILL constitutes an important point of reference for large-scale infrastructures who face the challenge of coordinating and managing collaborations between instrument scientists and user scientists, across a wide range of disciplines (D'Ippolito and Rüling 2019).

In the more than 50 years of its existence, the ILL has been subject to multiple upgrades and improvement programs, which have enabled a steady improvement of its instruments and neutron flux (Arai and Crawford 2009). Nevertheless, its continuation in the long run has regularly been called into question – most recently in the context of the establishment of the ESS in Lund, Sweden, which has been advertised by various stakeholders as a "like-for-like replacement" of the ILL (ESFRI 2016a: 7). However, although the construction of the ESS is ongoing – its user program is due to start in 2023 – influential actors within Europe's science policy arena have begun advocating a prolongation of ILL funding and operation.

The current intergovernmental agreement between France, Germany and the UK is valid until 2023, although the three countries have already expressed their commitment to apply for an extension until 2033 (ILL 2019). Whether governments will ultimately continue to support ILL or shift towards alternative neutron sources or technologies (a scenario similar to the transition/

switch from Fermilab to the Superconducting Super Collider in the United States (US); Hoddeson et al 2008) remains an open question. This fundamental uncertainty makes the case of the ILL ideal to gain an understanding of the efforts addressed towards keeping an RI alive.

Hallonsten and Heinze (2012) have discussed the institutional persistence of systems of national laboratories in Germany and the US since the mid-1950s, providing a rough categorization of their history into three periods: initiation, expansion and diversification. National laboratories constitute a specific type of research organization that, unlike universities, seem not to have undergone transformation in their sponsorship patterns and governance as a whole, and unlike corporate research centers, most national laboratories have not been reconfigured as interorganizational networks. This apparent macro stability contrasts with the fact that, although such organizations start as mission-oriented projects with a limited time horizon, most of them develop into multipurpose facilities with an apparently open-ended lifetime (Hallonsten and Heinze 2012: 451). The authors draw attention to an intricate relationship between organizational adaptation and persistence of the system within which RIs are situated, a focus that raises an important question about the type of persistence a large-scale infrastructure sustains against the time passing by. Subsequent contributions by the same authors have highlighted how institutional persistence of this takes place also at the level of the laboratory and facility (Hallonsten and Heinze 2016; Heinze and Hallonsten 2017; Heinze et al 2015).

As discussed in the introduction to this volume, the last three decades witnessed increased attention by the European Union (EU) toward RIs, defined as "facilities, resources and related services that are used by the scientific community to conduct top-level research in their respective fields" (European Council 2009: 4). Whilst a broad array of definitions have emerged throughout the years, there seems to be consensus on the strong relationship between the nature of these facilities and the science that they host. The introduction has once again illustrated different definitions of big science, emphasizing a shift in emphasis from large experiments in the early stages to small-scale experimental work conducted on very large instrumentation.

Our primary focus in this chapter is on the ILL's efforts over the past decade to maintain its leading position in neutron science (ESFRI 2016b; Hallonsten 2016a) and ensure operation until at least 2030, especially in the context of two major upgrade initiatives, the Millennium (2000–15) and Endurance (2016–23) Programmes. We do so by systematically analyzing archival data including annual reports from ILL (2005–18) and the European Strategy Forum on Research Infrastructures (ESFRI) (2004–12), the 2016 ESFRI Roadmap, and additional sources providing historical narratives about the ILL (e.g. Jacrot 2019). Our analysis focuses both on the reconstruction of the historical events surrounding the two upgrade initiatives and the narratives produced in the context of the ILL's extension efforts described below.

Our empirical evidence points out how the organizational efforts aimed at extending ILL's lifespan fall into three distinct, yet intertwined categories: efforts related to the ILL's "technical life", which regards material and technical aspects of managing the facility, such as the upgrade of instruments and infrastructure; efforts focusing on ILL's "social life", mainly through the preservation and renewal of its user community; and efforts connected with ILL's "political life", encompassing an array of institutional and political activities focused on maintaining the ILL aligned with shifts taking place within its broader science and policy context. For the sake of illustrating our findings, we have separated the presentation of these three domains in the sections that follow (Sections 3, 4 and 5, respectively). We will explore their connections and mutual dependencies in the concluding Section 6; there, we will theorize on how the temporal concept of lifespan applies to, and connects with, the technical, social and policy dimensions of RIs. Finally, we will also discuss wider policy implications from the perspective of the facilities (e.g. ongoing research collaborations, plans for future developments/upgrades of the facility), users and policymakers.

3. TECHNICAL LIFE: UPGRADING THE ILL REACTOR AND ITS SCIENTIFIC INSTRUMENTS

The duration of operations of an RI is limited by its technical lifespan. Synchrotron radiation sources and other particle accelerators, for example, are determined by their operation as concrete, physical installations. In addition, such facilities need to be run bearing in mind the regular updates that their scientific instruments require. The ultimate scope of this goes beyond maintaining the technical capacity of the instrumentation and includes pushing the frontiers of the science conducted within the facility to engage with new disciplines, user communities or experimental techniques, in order to maintain an RI "at the forefront of [...] science" (ILL 2007, 2008: 110).

The most visible aspect of keeping an RI alive thus comprises work dedicated to material and technical upgrades, including scientific instruments, sample and computing environments, facilities, etc. In the context of the ILL, technical upgrades and initiatives dedicated to improving the facility's operations represent a significant share of the organization's reporting – on average more than 20 percent of its annual reports refer to such initiatives.

Ensuring the operation of the ILL over an extended period required significant work and investment on a material and technical level, involving its scientific instruments and infrastructure as well as the neutron-generating reactor at the core of the facility. Since the year 2000, the ILL has undertaken two major modernization programs focusing on building or upgrading a large number of scientific instruments. The Millennium Programme, which lasted from 2000 until 2018, represented a total cost of almost €80 million (including more than €40 million of investments) and resulted, according to the ILL, in a 20-fold performance increase of the ILL's instruments (ILL 2019). The subsequent Endurance Programme was conceived in 2010 to pursue a further improvement of the ILL's instrument suite with the aim of "keeping the ILL at the forefront of neutron scattering for the next decade and beyond" (ILL 2013: 73). Following authorization by the ILL's Steering Committee in 2015, this program, which started in 2016 and is scheduled to last until 2023, involves further instrument upgrades, the creation of new instruments, as well as the renewal of neutron guides, improvements in sample environments and the development of new data analysis software. In addition to its ongoing efforts to modernize and upgrade its instruments and scientific infrastructure, the ILL also engaged in several initiatives to ensure the conformity of its neutron-generating reactor with changing nuclear safety standards, thus seeking to extend the lifespan of reactor operation, with a time horizon "well beyond the year 2030" (ILL 2008: 7).

The two dimensions of the ILL's material upgrades – the scientific instruments suite on the one hand and the neutron-generating reactor on the other – follow distinct logics. Ensuring the safe operation of the nuclear reactor and safeguarding its long-term operating horizon represent ongoing concerns, structured in time by a series of recurrent as well as unplanned events. Recurrent events include, for example, formal ten-year safety reviews under the responsibility of the French Nuclear Safety Authorities, or the renewal of contracts to ensure the reactor's supply in highly enriched nuclear fuel. Scheduled safety reviews reflect the evolution of nuclear safety regulation and have been key drivers of reactor building and facility upgrades to ensure compliance with changing safety standards, such as the ILL's 2002–2007 Refit Programme or the subsequent Key Reactor Components Programme. Unplanned events – in particular, during our study period, the 2011 Fukushima nuclear accident – demand further enhancements in terms of reactor safety, e.g. to ensure safe reactor operation under conditions of extreme flooding, which were addressed by the ILL through additional reactor-focused programs such as the Fukushima Programme, which required modifications costing more than €20 million (ILL 2015). In its reporting, the ILL systematically emphasizes the ingenuity of the facility's original reactor design, embodying "the foresight of those who originally conceived the reactor design, optimizing not only brightness but also robustness with regard to safety" (ILL 2012: 5), the reactor's high reliability in the past and the ILL's commitment towards maintaining high levels of reactor safety.

Taken together, reactor upgrades conform to a logic of safety and integrity as given norms against which the actual state and operation of the reactor can be measured. When it comes to upgrades of its instrument suite and of the user-facing aspects of its facilities, on the other hand, the ILL texts embrace a more active, goal-oriented rhetoric, emphasizing the organization's intent to set an agenda for the future development and possibilities of neutron science in Europe and beyond. It is on the level of its technical infrastructure and the performance of its instruments that the ILL constructs its role as a pace setter in neutron-based science, as expressed in the following text passage: "To maintain its status as leader in neutron science, the Institute has constantly upgraded its instruments, infrastructure and scientific equipment over the last 50 years. The latest modernization exercise – the Endurance Programme – will continue to develop instrumentation and support services with a view to maintaining the Institute's world-leading position for another decade at least" (ILL 2019: 7).

The Millennium and Endurance upgrade programs – both dedicated to improving scientific instruments together with infrastructure to enhance the ways in which users can conduct experiments, analyze data, etc. – represent large multiyear projects, conducted in multiple stages. The first stage of the Millennium Project, lasting from 2000 to 2008, involved upgrading eight of the ILL's scientific instruments and the building of six entirely new instruments (ILL 2009). Its second stage, initially scheduled to last from 2007 to 2014, comprised the upgrade of four additional instruments, construction of four new instruments, as well as upgrading three important neutron guides, improvement of sample environments and the development of new instrument control software (ILL 2011). From its first stage onwards, the program was presented as a success in terms of improved neutron flux and instrument performance, but also as an initiative that provided the ILL with a renewed sense of its role in advancing neutron-based science in Europe: "[The Millennium Programme] has created a renewed sense of purpose at the Institut and demonstrated that there still remains enormous untapped and cost-effective potential to be realized for the good of European science" (ILL 2006: 7). Upgrades to the *nuclear reactor* were required to respond to changes in nuclear safety regulations, for example as a response to the 2011 Fukushima nuclear accident. Guided by the ultimate goal of ensuring the safety of reactor operations, they represent a necessary condition for conducting scientific work at the ILL. Improving the *scientific instruments and infrastructure* on the other hand provided the opportunity to emphasize the ILL's position as a leading center in neutron science, including the development of new technologies for putting neutron scattering to use in a wide range of scientific disciplines. Whilst the reactor upgrades are typically associated with the fixed time horizon of safe future reactor operation, scientifically driven upgrades project the community towards open-ended progress and discovery as well as the possibility to reach the technologically feasible, allowing for instance to reap "astonishing gains, increasing the average detection rate across the instrument suite by a factor of more than 24 already, and transforming the scope and capacity of the science we support" (ILL 2013: 4).

The success of the Millennium Programme in enhancing the effectiveness of the ILL instruments served as a core argument to support the proposal of the subsequent Endurance upgrade program as an expression of interest of the scientific community:

> The Millennium Programme still has two years to run, but the process of renewal should not end there. It is vital to the health both of our Institute and the broader scientific community that we continue to explore and exploit the limits of what is possible in instrumentation. To this end, four years of consultation and planning will deliver in 2013 the case for the next wave of our upgrade Programme, christened "Endurance" to signal our aim to continue to lead for years to come. (ILL 2013: 4)

In terms of extending the facility's lifespan, the Millennium Programme was characterized as yet another "cycle of reincarnation" (ILL 2013: 4), and the choice of the name of its successor program, Endurance, also projected a sense of continuance and durability, especially with respect to the parallel advancement in the construction of the ESS in Sweden, which was initially conceived of by some policymakers as a replacement for the ILL (ESFRI 2016a: 7).

As the ESS project advanced, with the commissioning of first instruments expected for 2020 and first user experiments for 2023 (see above), European neutron user and science communities increased their efforts to stress the need to maintain the ILL in order to safeguard broad access to several top-level neutron sources, a necessary condition for European leadership in neutron science (see Section 6). In this context, upgrades of the ILL's scientific instruments were increasingly presented as a contribution not just to the ILL, but also to other facilities that would benefit from the knowledge and technologies developed in the context of the ILL upgrade projects: "Our achievements under the Millennium Programme are now set to cross frontiers: the benefits will be transferred to other neutron facilities through agreements on the supply of high-tech instrumentation – detectors, monochromators and polarization analysis devices" (ILL 2015: 79). This in particular included an emphasis on the ILL's "central role in preparing the scientific community for the European Spallation Source over the next ten years" (ILL 2016: 5). As we outline in more detail below, this new emphasis resonated with an overall shift in positioning the ILL from being a research infrastructure in its own right toward serving a large scientific community as "*the* European neutron source" (ILL 2019: 1, emphasis added).

4. SOCIAL LIFE: BUILDING AND MOBILIZING USER COMMUNITIES

Work directed at strengthening and broadening social ties through both interaction and rhetoric is strongly intertwined with work on the ILL's material upgrades. Large RIs such as the ILL often share a remit of serving international user communities of scientists who visit their facilities to conduct experimental work so that "the facilities do not produce any science themselves, their users do" (Hallonsten 2016b: 486).

The ILL itself is seen as a pioneer of this type of organization (Rush 2015). Each year, it hosts around 1500 international researchers conducting experiments on its instruments. Users are granted "beam time" on the basis of a competitive proposal selection process and have their travel and accommodation expenses covered by the ILL. Among other aspects, this latter point makes it easier for doctoral students and early career researchers to participate in experiments conducted at the ILL.

Portraying that "the real strength of ILL is human" (ILL 2013: 5), the ILL actively seeks to mobilize support by users and its broader scientific community. Our previous work on how collaborations unfold within an RI like the ILL has emphasized how the ILL's long-term success is associated with its ability to interconnect various forms of complex research collaborations ranging from standardized "instrument service" to unique projects carried out in a logic of "peer collaboration" to advance entire research domains (D'Ippolito and Rüling 2019).

The ILL leverages the relationship with users to emphasize a "user community that is second to none" (ILL 2012: 5) as one of its core strengths. Its institutional publications systematically highlight the critical role of users, as for example in the publication in 2017 of a "Souvenir Book" to celebrate the 50th anniversary of the ILL (ILL 2017), which in particular emphasizes the role the ILL played in scientists' early careers. The following user testimony from the "Souvenir Book" illustrates how the ILL leverages user experiences to construct its identity as an organization that connects its users "for the best science":

> [E]verybody was excited and proud to work and do research there, and everybody felt like they were part of a grander project […] It was the ILL, not just any neutron source offering services, but an advanced research centre with the special ability to connect people together […] I believe that openness, collaboration, inclusiveness and loyal competition were the perfect ingredients to the success of the ILL. It was its human factor gluing together the best technologies for the best science. (ILL 2017: 60)

To maintain their sense of connection to a community of ILL users, the ILL provides opportunities for participation in formal events such as the annual European User Meeting or the User Symposiums organized within the context of the ILL's upgrade programs. More informal interactions take place through "User Forums", which connect users on site with ILL directors and group leaders "to get feedback on [...] users' needs and [...] to be more responsive to emerging demands" (ILL 2006: 9). In September 2010, a three-day "ILL 2020Vision Meeting" in Grenoble played an important role in defining the contours of the upcoming Endurance Programme (ILL 2011: 9).

Claims of strong and stable ILL–user relationships combine narratives of personal user experience (often at early career stages), with reporting on highly productive scientific collaboration, and with the display of a shared sense among users and ILL scientists of forming an overall ILL-related community, embodied for example in regular user events. In the annual reports of the ILL, the sense of the ILL–user relationship is reinforced through frequent references to the critical role of users for the ILL's success. Extensive sections of the annual reports (accounting for roughly 50 percent of the reports) delve into "Scientific Highlights", featuring experimental results across disciplines in the form of short, academic articles co-authored by users and ILL scientists, and each annual report also includes a two-page photo section representing "Happy Users" during their stay at the ILL.

In addition to the material upgrades aimed at providing *individual* users with advanced scientific instrumentation, and of its active engagement in constructing a shared sense of user *community*, the ILL engages in a range of more formal, interorganizational networks among neutron facilities and research infrastructures more broadly. These include, for example, its active role in the ESFRI, especially its working group on Neutron Landscape established in 2014, as well as the coordination of SINE2020, a multi-year EU-funded project (2015–19, €16 million budget) involving 18 European neutron sources and research laboratories to develop new and improved technologies and services for scientific and industrial users across Europe.

As we will illustrate in the following section, it is the construction and mobilization of these formal and informal networks that enable the ILL to establish itself as a key actor within the European science policy.

5. POLITICAL LIFE: POSITIONING THE ILL IN ITS POLICY CONTEXT

The sections above have highlighted how the material component of managing RIs (i.e. instruments, infrastructure, etc.) is strictly intertwined with the science supporting it (Section 3) as well as the user communities that the shared facility ought to serve (Section 4). Managing such a complex and intricate system is becoming ever more challenging, especially as international collaboration is on the rise (Lee et al 2015) and tends to be more interdisciplinary in nature (Stephan 2012). Despite the relevance of RIs in undertaking public science, efforts to unpack how these infrastructures may need to be managed in the longer run vis-à-vis the political and financial difficulties that they may encounter remain scant (Lossau 2012).

A substantial portion of the literature on the history of major laboratories has discussed how the technical obsolescence of large-scale facilities has often prevented additional investments to either maintain or continue to keep them in operation. For instance, Westfall (2010) illustrated how the US Argonne National Laboratory depended very much on the fate of the Intense Pulsed Neutron Source (IPNS). Since the lab was of quintessential importance for the surrounding scientific ecosystem, Argonne had to prevail in contests with other laboratories and obtain large projects not only to gain funding, but above all to justify its existence at a time when the national laboratory environment had grown harsh and inhospitable (Westfall 2010: 361). What enables such large-scale facilities to keep their operations alive for as long as possible? We contend that historical legacy and policy work play a major role in encouraging investments on these sites.

According to its own narrative, the ILL has played a key part in the advancement of neutron science over the last 50 years. The texts we analyzed suggest that both political and scientific stakeholders were determined to make the ILL a success from the beginning. Since the project was forged and promoted by the leading scientists of the time, it was believed that the ILL should receive support not only because of its scientific potential but also because of its political significance, when Europe was undergoing a process of rebuilding just 20 years after a traumatic conflict (ILL 2017: 2; Cramer, ch 3 in this volume). This historical legacy has enabled the ILL to maintain its projects through constant references to its own scientific success, in turn supporting the mobilization of political support.

Individual-level, project, scientific field and user community dimensions of relational work conducted by the ILL are strongly embedded within the discursive construction of its distinctive European identity and an emphasis on the contribution of the ILL to scientific activity in Europe more generally. The ILL's European policy focus was reinforced over the past two decades culminating in the deliberate branding of the ILL as "the European neutron source" on the cover page of its 2018 annual report (ILL 2019: 1). In terms of ensuring the longer-term operation of the ILL, the construction and maintenance of relational networks on various levels and their interweaving are critical for gaining support by science policymakers. It is telling in this context that one of the testimonies published by the ILL regarding its career-changing influence on young researchers ends with the assertion that "[t]he ILL and its culture have to be protected because they belong to all European researchers" (ILL 2018: 60).

Positioning itself as the world leading research center for neutron science, the ILL presents itself as a political voice of relevance both at the national and international levels, in particular within the European policy towards neutron science. It has historically tried to gain political weight; at the national level, the interest in shaping policy has primarily been associated with decision making related to the safety of neutron infrastructures. In 2005, the ILL led a joint action (on behalf of the three European institutes on the common site, the European Molecular Biology Laboratory, the ESRF and the ILL) to be recognized as a partner by the French local authorities when defining infrastructure policies. The benefit of this action was indeed immediate: the ILL could rely on the special national French funding (Contrat de Plan Etat-Régions) to undertake major infrastructure projects (Jacrot 2019: 236).

Continuous investments on maintaining or renewing its infrastructures, such as the Millennium Programme or the Endurance Programme, have guaranteed the ILL a meaningful political voice also within the European science arena, although this political support is not to be taken for granted. Whilst beyond the temporal focus of this chapter, the substantial changes to the reactor required in the 1990s provide an example of how the partners' consensus was not always as obvious. The required long-term reactor shutdown and high investment required had a destabilizing effect at a time partners were also involved in establishing the ESRF or considering other sources to take over (e.g. ISIS in the UK or the German reactors in Jülich and Berlin). Political support comes hand in hand with financial support: in the case of the long 1990s reactor shutdown mentioned above, the ILL gained the partners' consensus only because its management agreed that the reactor renovation would be performed within the ILL's annual budget, without modification of previously established partner contributions. It is worth pointing out that, although the three associate countries (France, Germany and the UK) cover most of the annual budget (79 percent in 2015), the ILL must reach a consensus among *all* scientific member countries (who are represented in the ILL's Steering Committee and Scientific Council) in order to gain political weight across the international landscape.

It seems as an obvious assertion that undertaking technical advancements leads to improved cost effectiveness – however, the emphasis on this dimension has not been a constant throughout the period we are observing. We found that, in the period between 2005 and 2010, the organization was highlighting how it had been successful at reducing the number of unscheduled downtime days and improving the facility's instrument performance by a factor of 20, mainly through enhanced neutron flux, before and after the Millennium Programme (ILL 2019). As of 2010 onwards, the ILL's discourse more strongly emphasized forging an organization that could represent the cornerstone of neutron science in Europe, engendering a positive impact on the broader scientific community. As emphasized by Jacrot (2019: 229), the Millennium and (current) Endurance Programmes, whilst being successful and cost-effective investments, are contributing to maintain the ILL's leading position among neutron scattering centers worldwide. The former, launched in 2000, had the remit of enabling the ILL to lead cutting-edge research in neutron science; nearly two decades later, the Endurance Programme is a testimony of how the ILL, after some 50 years of forefront neutron research, is continuing to equip itself for the purpose. Therefore, technical upgrades are above all instrumental to making the ILL a unique European organization at the forefront of science, which not only maintains a live neutron science on a global scale, but also fosters Europe's leadership in neutron science. This is mirrored in the ILL's annual reports, for example in statements that the ILL constitutes "an exceptional centre of excellence and a fine example of successful cooperation in Europe" (ILL 2009: 3).

Based on our reading of the archival material, a leading position such as the one held by the ILL is of benefit to various audiences, which in itself reinforces the reasons for keeping the organization in operation: the user base first and foremost. As discussed in our recent contribution (D'Ippolito and Rüling 2019: 1283), RIs like the ILL have to tackle two fundamental challenges, that is: growing the organization's user base and, at the same time, accommodating expert users to ensure instrument development at the leading edge of science. Instrument upgrades and new infrastructure investments have aimed at addressing these challenges: on the one hand, the interest of the wider scientific community has been leveraged upon (Elzinga 2012); on the other hand, the newly created supported facilities enabled both ILL scientists and user communities to adapt to the changing demands of science (Hallonsten 2016a; Hallonsten and Heinze 2016). More practically, this has meant that users with varying levels of experience, when visiting the ILL, are supported with teams of researchers, including the local contact, as well as a range of services that complement the "simple" offering of neutron beams (cf. Cramer et al, ch 1 in this volume).

Driven by the user community, the leading position held by the ILL is presented as beneficial to the wider scientific community within Europe. Today the ILL carries the major load for the overall European delivery of neutron instrument days, supported by the ISIS Neutron and Muon Neutron Source (UK), the Open Access Spallation Source SINQ (Switzerland), Forschungs-Neutronenquelle Heinz Maier-Leibnitz (FRM II, Germany) and some smaller sources in Europe. With an ongoing development of neutron scattering in the materials and life sciences, it is the urge and priority of not only the ILL but also of other neutron sources and funding agencies to nurture and protect the ILL's leading position (ILL 2017: 99). In order to develop a better understanding of the EU landscape of neutron research infrastructures, especially given the planned closure of many reactor-based sources in 2014–16 and the opening of the ESS in the 2020s, the ESFRI set up an Expert Group on Neutrons in 2004. Focusing on the implications in terms of capacity and capability of neutron science in Europe, the group identified a series of priorities. Among many, two are worthy of mention: "(i) those who fund Europe's neutron sources must begin the process of taking decisions Europe-wide rather than unilaterally; (ii) that the user community makes its voice heard" (ILL 2017: 98). In other words, although recognizing the ILL as holding a leading position, it is ever more important to adopt a European-wide perspective, of which the users are a key part.

When referring to the wider scientific community, we cannot neglect the new (currently under construction) neutron source in Europe, the ESS. In the words of the ILL's directors, "the future of ILL holds the key to maintaining Europe's lead in this field" (ILL 2017: 99). At the same time, a note by Colin Carlile (former director of both the ILL, 2001–2006, and the ESS, 2007–12) in ESFRI's report on neutron scattering facilities in Europe pointed out how the eventual sunset of the ILL must be handled with openness and awareness of its consequences on all levels, and regardless of how the establishment of the ESS unfolds (ESFRI 2016a). Both staff and users had mixed feelings about planning for the ESS as the successor of the ILL, however, it has become apparent that without the future promise of the ESS, the ILL would risk losing a number of national communities. It is interesting to observe how the very first director of the ILL, Bernard Jacrot, admitted that "ESS would provide unparalleled possibilities for European scientists" (Jacrot 2019: 202). In other words, the fates of the two institutes are very much interdependent: on the one hand, the ILL leads on neutron science research worldwide, yet its reactor does not have an infinite life, and it is not clear how long it will be in operation after 2030. On the other hand, the ESS will thrive on what the ILL has built over the years both in terms of educating a large user community and benefiting from expert scientists who have in the past worked for, or been regular visitors at, the ILL and currently have joined the ESS. An important aspect that is not explicitly addressed in our empirical material relates to the question of how realistic the dismantling of the ILL reactor actually is. Decommissioning a nuclear reactor is both technically difficult and extremely costly, which would constitute yet another reason to keep the reactor alive.

6. DISCUSSION AND CONCLUDING REMARKS

The ILL is used here as a case of an RI that leads research in neutron science. Our chapter connected with existing contributions examining the capability of RIs to adapt and renew their institutional and organizational remit over time through different forms of institutionalization (i.e. physical infrastructures, scientific fields, politics) (Hallonsten 2016a). The case of the ILL resembles an example of endurance whereby not only the ILL, but also the aspirations of the entire European neutron community, are at stake. The latest reactor upgrades should ensure safe reactor operation until at least 2030, yet as the ILL's reactor is ageing, there is huge uncertainty about what will happen after this time. The future scenario complicates further given the prospects of the ESS that, despite being complementary to the remit and scientific orientation of the ILL, may at some point fully substitute the Grenoble-based facility.

Our empirical evidence suggested that the ILL and its wider community, in particular other member countries from Europe, strive to keep the reactor running. We observed a keen interest of the ILL's management to make sure that the instruments enable the execution of cutting-edge experiments as well as constantly improving the overall ILL user experience, for example by developing new computing, data management and archival infrastructure to cope with the exponential growth of data generated by ever higher resolution instruments. Efforts have gone as far as developing new control software environments standardizing to facilitate experimental set-up across instruments. We observed a continuous commitment on the organization's side aimed at: reinstating the role of the ILL within the neutron science landscape; strengthening the organization's ties with scientific disciplines and user communities with whom ILL did not interact before (e.g. collaborative initiatives with the European Molecular Biology Laboratory (EMBL)); or harvesting on connections with other neutron sources (e.g. by positioning the ILL as complementary to the ESS).

This chapter has focused on three main dimensions that characterize the efforts of the ILL to extend its lifespan. First, it is about making sure that the reactor is running safely and without disruptions. Clearly, upgrading the instruments is an important component of connecting with the scientific community, but the instruments would be meaningless if the reactor would not be generating neutrons for them. Second, it is the reliability of the ILL's reactor source in conjunction with the performance of its instrument suite that forms the basis for the ILL's ability to strengthen and broaden its user community. Finally, yet importantly, political and financial support mediated by the ILL's scientific community backing constitutes the foundations for guaranteeing a long-term perspective to the facility. The ILL is flexible and open-ended, entailing that it can adapt to changing political and scientific demands. In a wider sense, the ILL's scientific community is to be intended not only for the visiting user base, but also the scientific members and other, potentially competing sources; although the RI may be leading the science through cutting-edge research, it is fundamental to gain consensus (i.e. legitimacy) by engaging with, and contributing to, the proficient running of other organizations. Furthermore, with the increasing demand for science to be relevant for and have an impact on society, large RIs like the ILL have started to draw light on how their range of methods is beneficial for audiences beyond academia, also facilitating the user communities and scientific staff to adapt to it.

We recently argued that the co-existence of multiple collaboration types within the same organization, driven by both instrument scientists' decisions and the structures and resources provided at the organizational level, serves as an important ingredient for the long-term success of large RIs (D'Ippolito and Rüling 2019). Building on this, the current chapter illustrated how the ability of a facility like the ILL to extend its lifespan depends on the organization's skill to leverage on and balance the three dimensions named above. In different terms: users can be "educated", but if the RI does not make the effort to constantly upgrade its instruments and infrastructure, then users may decide to transfer their experimental activities to another source; by the same token, leading-edge instruments alone are no guarantee of a smooth run for a facility. Instrument scientists and the organization more broadly must entertain caring relationships with the user base, making sure visiting users are aware of the upgrades and the possibilities these may generate for neutron science. Last but not least, if the organization cares about the users and yet, this interest is not translated into political support, then the RI may not be able to sustain the funding necessary for its operation and technical upgrades. Therefore, it is essential that the three parts come together: RIs need to leverage on the material aspect and technical performance of instruments and infrastructure to attract, maintain and interrelate users; consensus emerging from a strong user community will provide the RI with a political voice and support of its objective to extend its life in the long run, producing what Hallonsten and Heinze (2012) define as "institutional persistence".

For the aims of this research, we have relied primarily on archival sources and additional academic references. Future research may explore and observe in situ how the three forms of work described above interrelate and shape collaborations within and around RIs, ultimately contributing to maintaining both the RI and its scientific community on the edge of breakthrough science.

What the future holds for the ILL is of course all to discover. We could envision the mobilization of policies that further extend the technical operating horizon of the nuclear reactor upon meeting maintenance requirements or – on the contrary – the withdrawal of political and financial support by some of its scientific member countries. Admittedly so, the latter scenario is less likely to happen given that, in Spring 2018, the three ILL associates have declared "their intention [...] to sign a sixth protocol, securing its operation for another ten-year period: 2024–2033" (ILL 2019). Regardless of future developments, the case of the ILL has provided insight on the importance for an RI to maintain its technical and scientific edge as well as strengthening and politically leveraging its user community. The empirical discussion sheds light on how RIs may respond to changing circumstances of "their" scientific world and beyond via different forms of institutional resistance, that is, physical, social and political.

ACKNOWLEDGMENTS

We thank Dr Giovanna Cicognani, head of the User Office at Institut Laue-Langevin, for her continuous support. The opinions expressed in this chapter are those of the authors. The usual caveats apply.

REFERENCES

Arai M and K Crawford (2009) Neutron sources and facilities. In I S Anderson, R L McGreevy and H Z Bilheux (eds) *Neutron imaging and applications: A reference for the imaging community*. Springer Science and Business Media, pp 13–30.

Banks M (2014) CERN kicks off plans for LHC successor. *Physicsworld* February 6.

Brumfiel G (2004) Views collide over fate of accelerator. *Nature* **442** (7103).

Crease R P (2001) Anxious history: The High Flux Beam Reactor and Brookhaven National Laboratory. *Studies in History and Philosophy of Science* **32** (1): 41–56.

Crease R P (2005) Quenched! The ISABELLE saga II. *Physics in Perspective* **7** (4): 404–52.

D'Ippolito B and C-C Rüling (2019) Research collaboration in large scale Research Infrastructures: Collaboration types and policy implications. *Research Policy* **48** (5): 1282–96.

Elzinga A (2012) Features of the current science policy regime: Viewed in historical perspective. *Science and Public Policy* **39** (4): 416–28.

ESFRI (2016a) Neutron scattering facilities in Europe: Present status and future perspectives. In C Carlile and C Petrillo (eds) *ESFRI Scripta. Vol. 1: Physical sciences and engineering strategy working group*, pp 1–133.

ESFRI (2016b) Strategy report on Research Infrastructures: Roadmap 2016. In Science and Technology Facilities Council (ed) *European Strategy Forum on Research Infrastructures*, pp 1–206.

European Council (2009) Council regulation No 723/2009 of 25 June 2009 on the Community legal framework for a European Research Infrastructure Consortium (ERIC) [2009] OJ L 206/1, as later amended by Council Regulation (EC) No 1261/2013 of 2 December 2013 [2013] OJ L 326/1.

Hallonsten O (2009) *Small science on big machines: Politics and practices of synchrotron radiation laboratories*. Doctoral dissertation, Lund University.

Hallonsten O (2015) Unpreparedness and risk in Big Science policy: Sweden and the European Spallation Source. *Science and Public Policy* **42** (3): 415–26.

Hallonsten O (2016a) *Big Science transformed: Science, politics and organization in Europe and the United States*. Palgrave Macmillan.

Hallonsten O (2016b) Use and productivity of contemporary, multidisciplinary Big Science. *Research Evaluation* **25** (4): 486–95.

Hallonsten O and T Heinze (2012) Institutional persistence through gradual organizational adaptation: Analysis of national laboratories in the USA and Germany. *Science and Public Policy* **39** (4): 450–63.
Hallonsten O and T Heinze (2015) From particle physics to photon science: Multi-dimensional and multi-level renewal at DESY and SLAC. *Science and Public Policy* **40** (5): 591–603.
Hallonsten O and T Heinze (2016) "Preservation of the laboratory is not a mission": Gradual organizational renewal in national laboratories in Germany and the USA. In T Heinze and R Münch (eds) *Innovation in science and organizational renewal.* Palgrave, pp 117–45.
Heinze T and O Hallonsten (2017) The reinvention of the SLAC National Accelerator Laboratory, 1992–2012. *History and Technology* **33** (3): 1–33.
Heinze T, O Hallonsten and S Heinecke (2015) From periphery to center: Synchrotron radiation at DESY, Part I: 1962–1977. *Historical Studies in the Natural Sciences* **45** (3): 447–92.
Hoddeson L, A W Kolb and C Westfall (2008) *Fermilab: Physics, the frontier and megascience.* University of Chicago Press.
Holmberg G (2013) The momentum of maturity: What to do with ageing Big Science facilities. In T Kaiserfeld and T O'Dell (eds) *Legitimizing ESS: Big Science as collaboration across boundaries*. Nordic Academic Press.
ILL (2006) *Annual report 2005, Institut Laue-Langevin.*
ILL (2007) *Annual report 2006, Institut Laue-Langevin.*
ILL (2008) *Annual report 2007, Institut Laue-Langevin.*
ILL (2009) *Annual report 2008, Institut Laue-Langevin.*
ILL (2011) *Annual report 2010, Institut Laue-Langevin.*
ILL (2012) *Annual report 2011, Institut Laue-Langevin.*
ILL (2013) *Annual report 2012, Institut Laue-Langevin.*
ILL (2015) *Annual report 2014, Institut Laue-Langevin.*
ILL (2016) *Annual report 2015, Institut Laue-Langevin.*
ILL (2017) *1967–2017 | The Institut Laue-Langevin: 50 years of service to science and society.* Neutrons for Society.
ILL (2018) *Annual report 2017, Institut Laue-Langevin.*
ILL (2019) *Annual report 2018, Institut Laue-Langevin.*
Jacrot B (2019) *Neutrons for science: The story of the first firty years of the Institut Laue-Langevin: A very successful international cooperation.* EDP Sciences.
Lee Y-N, J P Walsh and J Wang (2015) Creativity in scientific teams: Unpacking novelty and impact. *Research Policy* **44** (3): 684–97.
Lossau N (2012) An overview of research infrastructures in Europe – and recommendations to LIBER. *Liber Quarterly* **21** (3/4): 313–29.
Lozano S, X-P Rodriguez and A Arenas (2014) Atapuerca: Evolution of scientific collaboration in an emergent large-scale research infrastructure. *Scientometrics* **98** (2): 1505–20.
Rush J J (2015) US neutron facility development in the last half-century: A cautionary tale. *Physics in Perspective* **17** (2): 135–55.
Stephan P E (2012) *How economics shapes science*. Harvard University Press.
Weinberg A M (1967) *Reflections on Big Science.* MIT Press.
Westfall C (2008) Retooling for the future: Launching the Advanced Light Source at Lawrence's Laboratory 1980–1986. *Historical Studies in the Natural Sciences* **38** (4): 569–609.

Westfall C (2010) Surviving to tell the tale: Argonne's Intense Pulsed Neutron Source from an ecosystem perspective. *Historical Studies in the Natural Sciences* **40** (3): 350–98.

12. Big Science and Research Infrastructures in Europe: Conclusions and outlook

Olof Hallonsten and Katharina C. Cramer

This book builds on the assumption that there is more than the widely used but still very vague category of Big Science to describe and investigate particularly large research projects and facilities. The introductory chapter identified four categories, Big Science and Research Infrastructures (RIs, capitalized), and big science and research infrastructures (non-capitalized), of which the former two were identified as the core concerns of this book. Drawing inspiration from previous work, the two categories were framed in a way that hopefully helps scholars to navigate through the different kinds and dynamics of large-scale research in historical and contemporary perspectives. The chapters of this book, written by scholars of different fields with complementary interests, have shown the way: The book brandishes a remarkable collection of empirically well-found studies that bring the study of Big Science and RIs in Europe up to speed with developments in the most recent decades. With regards to conceptual development, however, the progress made in this book is less evidently clear.

Generally speaking, big science (non-capitalized) is a very broad description of science grown big in several ways – costlier, more densely populated and with a dramatically larger output (Price 1986/63), perhaps organized in industry-like manners (Weinberg 1961), but also more exposed to political and economic realities and with bigger expectations placed on it from society. Especially with regard to the latter two aspects, there is a particular connection to Big Science (capitalized), namely that this category does not only describe the general growth of science but that its conceptual meaning and use in scholarly research points to particular political contexts and logics of science. We know from previous work that Big Science was born at the end of World War II and grew important in the early Cold War as a particular form of science that used large-scale instrumentation and that was closely tied to the military and political logic of the superpower competition between the Soviet Union and the United States. This situation changed in recent decades, as did the socio-economic and political contexts leading to a reframed understanding of Big Science such as "Big Science Transformed" or "New Big Science" (Hallonsten 2016a; Crease and Westfall 2017).

There is, moreover, particular kinship between Big Science and RIs (both capitalized), to the extent that the politics of the two have seemed to significantly overlap in Europe since the early twenty-first century. As discussed in several of the chapters in this book, RIs is very much a political phenomenon, and the concept itself is predominantly a rhetorical device and thus a discursive policy tool that especially policymakers and bureaucrats in the European Union (EU) have pulled out of their sleeves in recent decades in order to mobilize resources and not least attention around an evidently important feature of current science and innovation systems. As pointed out in the introductory chapter, the definition of the European Strategy Forum on Research Infrastructures (ESFRI) is political and not technical or analytical, and the process of ESFRI's "consecration" of projects is one of political priority setting and negotiation rather than adherence to strict and predefined criteria. Expectations on what RIs can contribute to the development of the European knowledge-based economy are high. Still, the collections of instruments, data repositories, vessels and observatories of various kinds that are found on the roadmaps of ESFRI and of national agencies and bodies are extremely broad and diverse. This means that it is hard to find or develop any definition of RIs that does not either become entirely political or collapse into a wide and generic definition of research infrastructures (non-capitalized), i.e. simply material or organizational resources that can be utilized to do research, which includes virtually everything.

There was, hence, good reason to make a thorough effort to connect the strand of literature and scholarly work on Big Science to the policy hype around RIs in Europe today and in recent decades, and to discuss the overlaps between Big Science and RIs. However, in this sense, this book also suffers from exactly the same conceptual weakness or flaw as its predecessor volume from 1992, namely *Big Science: The growth of large-scale research*, edited by Peter Galison and Bruce Hevly. In the concluding chapter of that book, Bruce Hevly concludes that "even after hundreds of pages of text, 'big science' itself remains an elusive term" (Hevly 1992: 355). In one sense, in this book, we are worse off because we have two concepts – Big Science and RIs – that, after yet another few hundreds of pages of text, both continue to remain elusive. We might therefore just as well be clear on this point, straight away: Those who want concise definitions should look elsewhere. We – the 14 authors of this book – did not invent the concepts Big Science or RIs. If it was up to us with ambitions of conceptual and analytical stringency, we would probably have used other concepts and terms. But then, on the other hand, our work would most likely not have found its proper audience. This audience, you the reader, will have to do with other learnings, beyond conceptual definitions.

In this regard, the concept of Big Science quite clearly remains the most prominent scholarly approach towards large-scale research. The study of Big Science has taken several steps forward in recent years, and several of the contributions to this book add to this progression. The original rationale of Big Science, characterized by huge governmentally sponsored and mission-oriented research and development programs and projects, but also large equipment, exemplified especially by particle accelerators for particle physics experiments and nuclear reactors that were put in place to enable scientific discovery, is long gone. Nonetheless, Katharina Cramer's contribution to this book highlights that Big Science continues to play an important role, and that there are good reasons for the continued study of Big Science. In Chapter 3, she provides a historical overview of the founding phases of major Big Science projects in Europe ranging from the European Organization for Nuclear Research (CERN) in the 1950s to the European X-Ray Free-Electron Laser (XFEL) and the Facility for Antiproton and Ion Research (FAIR) in the 2010s, arguing that one aspect of the enduring role and significance of Big Science in Europe is its important role in times of political crisis and disarray that remain intact. Therefore, regardless of what we think of the concept itself, there is ample reason for continued attention to it.

However, the large body of scholarly literature on Big Science and emerging research on RIs is scattered across disciplines. Bringing clarity into this jungle was the main concern of Nicolas Rüffin's contribution to this book. In his chapter, he provided a splendid overview of the different research methods that were used to investigate the politics of Big Science and RIs, bringing together a comprehensive literature review and a collection and analysis of different methods and methodological approaches that complements and adds to classic readings such as Traweek (1988), Capshew and Rader (1992) and Galison and Hevly (1992).

In combination, these two contributions from Katharina Cramer and Nicolas Rüffin therefore highlight quite neatly how and to what extent Big Science, as an empirical phenomenon and a scholarly perspective, still has its role to play. In this regard, RIs is a different thing; until very recently, hardly any scholarly attention was paid to this concept at all, which means that this book's engagement with RIs as emerging phenomena, particularly in the European context, amounts to significant progress in the area. Considering their current role in policymaking, RIs have ostensibly become as important as Big Science was several decades ago – if not more – and as noted above, their role as assets in political rhetoric on the EU level is huge, although it remains to be seen how and to what extent this political-rhetorical significance will be matched by a demonstrable and significant contribution of RIs of all different flavors to the

European Research Area objectives of making Europe the most competitive knowledge-based economy in the world.

The three chapters by Inga Ulnicane (ch 4), Isabel Bolliger and Alexandra Griffiths (ch 5) and Maria Moskovko (ch 6) illustrate the political hype around RIs in Europe, but they also show that RIs are empirical realities, although perhaps not in the way historians and social scientists with knowledge in the history of Big Science would expect. The enormous variety of projects taken up on ESFRI's roadmaps and granted European Research Infrastructure Consortium (ERIC) status would make any empirically and theoretically well-behaved social scientist confused (see e.g. Hallonsten 2020). This leaves the scholarly study of RIs in Europe in a situation where it seems anything can be an RI and an RI can be anything. But as the chapters by Bolliger and Griffiths and Moskovko show, RIs are real and palpable empirical phenomena well worth deeper studies from the perspectives of political science, organizational sociology and international relations. The roadmapping processes of ESFRI and several European countries, and the transformation of existing and planned projects into ERICs, are promising topics of study in their own right. Similarly, the political-rhetorical hype around RIs on the EU level is itself understudied, and the journey of the concept of RIs into mainstream EU politics is in need of further analysis. In this regard, Inga Ulnicane's contribution to this book (ch 4) provides a comprehensive analysis of the current situation and a competent connection to classic readings in EU studies (e.g. Schimmelfennig 2016). Taken together, the contributions from Ulnicane, Moskovko as well as Bolliger and Griffiths can be regarded as a formidable, complementary overview of what the role of Big Science and RIs in current EU policy and politics is. Each of the three chapters add important pieces to the puzzling question of how, why and to what extent the role of the EU in large-scale research increased or intensified in recent decades.

Similarly, the use of the concept of RIs as a policy tool in national contexts and in disciplinary areas beyond the natural sciences is also in need of analysis, and Thomas Franssen has made an important contribution there. Not only showing how RI funding became a policy tool in a national context (the Netherlands) that mirrored the expectations and agendas of policymakers, he also highlights this development in the field of humanities and social sciences, which have their own RIs far from large accelerators and observatories – an underrepresented topic in studies so far (although exceptions exist) with their own governance, financing and management challenges.

But also the concept of research infrastructures (non-capitalized) gives rise to interesting theoretical questions. Treated quite simply as material and organizational resources for research, research infrastructures could presumably take on vital roles in current science and innovation systems – roles that require in-depth case studies to reveal the nature and extents thereof.

The chapter by Hallonsten, Eriksson and Collsiöö (ch 8) is exactly that; a deep case study of the involvement of Swedish nuclear energy research in the Halden Reactor Project in Norway. It becomes clear that the role of this research infrastructure is absolutely vital for large communities in the specific fields it serves, but that this role is difficult to capture and justify with contemporary commonplace indicators. The policy implications are significant: If the political hype around RIs in Europe is not matched by excellence and significance demonstrable with current metrics, a mismatch will be the result, and discontent might spread. Further studies on the topic of how to evaluate and acknowledge the role of RIs in innovation systems are clearly needed. Similarly, and related, the crucial issue of access to RIs is in dire need of scholarly research that goes beyond popular notions of fee-for-service use of RIs as if they were assets on a market where the business logic dominates. The chapter by Andrew Williams and Jean-Christophe Mauduit (ch 9) complements contributions made to the topic of access to RIs by others (e.g. McCray 2000; Hallonsten 2016b), adding several new insights as well as openings for important and interesting future research: The fundamental considerations of who gets access, and why, are clearly linked both to changing science policy rationales and the economization of science together with demands of relevance, efficiency and accountability. Moreover, through its case study, the chapter provides an interesting illustration of how ground-based astronomy positions itself within these new contexts.

Big Science is costly, RIs need not to be. The social study of big infrastructure projects outside of science and research has so far been disjointed from the social study of Big Science, but with the chapter by Hallonsten (ch 10), the theoretical advances of economic geographer Bent Flyvbjerg have now been tried on Big Science, and the analytical consequences sorted out: At least in part, the "iron law of megaprojects" holds also for Big Science. Not for all RIs, however, and this is important also as a reminder of the fact that the overlap between the two categories that correspond to the two concepts in the title of this book – Big Science and RIs – is not complete. All Big Science is not RIs, and vice versa.

Bigness breeds inertia and stability, but previous studies have shown that adaptability and the capacity of flexible mobilization of resources in different realms is necessary for the long-term survival of Big Science facilities (Hallonsten and Heinze 2012, 2013, 2016). In this book, these findings are taken further, in the first ever in-depth case study of the Institut Laue-Langevin (ILL) in Grenoble, generally credited with having consolidated and lifted trans-European neutron scattering to new heights from the 1970s onwards. The chapter by Beatrice D'Ippolito and Charles-Clemens Rüling (ch 11) shows where such success comes from. While the foundational political work necessary for the creation of Big Science facilities and costly and complex

RIs was the topic of Katharina Cramer's chapter (ch 3), the political work of keeping Big Science facilities and RIs alive and successful in the long run is the topic of Chapter 11, and it is clear from their analysis that such work is (almost) as multifaceted and varied as the technical and scientific complexity of the facilities themselves.

This book is a brand new collection of chapters dealing with previously known and unknown features of Big Science, launching the study of RIs as a promising line of inquiry in multidisciplinary social science, and with predominantly young authors. A generational shift is taking place by the publication of this book, amending and complementing previous analyses of Big Science – among the most prominent shall be mentioned Galison and Hevly (1992), Krige (2003, 2006), Westfall (2008), Crease (1999) and Hoddeson et al (2008) – but also broadening the disciplinary basis for such studies from predominantly history to sociology, political science, international relations and innovation studies. This generational shift and broadening of the disciplinary basis might alienate some readers who are well acquainted with the field in its previous form. Our intention is not, however, to either repudiate past contributions or dogmatically replace their perspectives and viewpoints with new ones. The purpose of this book is to expand the view, to augment the study of Big Science with new perspectives, to launch a scholarly effort of coming to terms with the new and highly politicized phenomenon of RIs, and to explore how the politicization of Big Science, which arguably never weakened but took new shape, acts out in the twenty-first century. The topics are far broader than what the chapters in a volume like this can encompass. With this contribution, we have hopefully ensured that the tradition continues and that several new lines of research are opened up.

ACKNOWLEDGMENTS

We, the editors, would like to thank all authors for their great efforts of contributing to this book. We would also like to sincerely thank Lina Rönndahl, whose talent, esprit and help in getting this book project on track cannot be overestimated.

REFERENCES

Capshew J H and K A Rader (1992) Big science: Price to the present. *Osiris* 2nd series 7: 3–25.

Crease R (1999) *Making physics: A biography of Brookhaven National Laboratory, 1946–1972*. University of Chicago Press.

Crease R and C Westfall (2017) The new Big Science. *Physics Today* **69** (5): 30–6.

Galison P and B Hevly (eds) (1992) *Big science: The growth of large-scale research*. Stanford University Press.

Hallonsten O (2016a) *Big Science transformed: Science, politics and organization in Europe and the United States*. Palgrave Macmillan.
Hallonsten O (2016b) Use and productivity of contemporary, multidisciplinary Big Science. *Research Evaluation* **25** (4): 486–95.
Hallonsten O (2020) Research Infrastructures in Europe: The hype and the field. *European Review* **28** (4): 617–35.
Hallonsten O and T Heinze (2012) Institutional persistence through gradual adaptation: Analysis of national laboratories in the USA and Germany. *Science and Public Policy* **39**: 450–63.
Hallonsten O and T Heinze (2013) From particle physics to photon science: Multidimensional and multilevel renewal at DESY and SLAC. *Science and Public Policy* **40**: 591–603.
Hallonsten O and T Heinze (2016) "Preservation of the laboratory is not a mission": Gradual organizational renewal in national laboratories in Germany and the United States. In T Heinze and R Münch (eds) *Innovation in science and organizational renewal: Historical and sociological perspectives*. Palgrave Macmillan, pp 117–45.
Hevly B (1992) Introduction: The many faces of big science. In P Galison and B Hevly (eds) *Big Science: The growth of large-scale research*. Stanford University Press, pp 1–7.
Hoddeson L, A Kolb and C Westfall (2008) *Fermilab: Physics, the frontier and megascience*. University of Chicago Press.
Krige J (2003) The politics of European scientific collaboration. In J Krige and D Pestre (eds) *Companion to science in the twentieth century*. Routledge, pp 897–918.
Krige J (2006) *American hegemony and the postwar reconstruction of science in Europe*. MIT Press.
McCray W P (2000) Large telescopes and the moral economy of recent astronomy. *Social Studies of Science* **30** (5): 685–711.
Price D J dS (1986/1963) *Little science, big science ... and beyond*. Columbia University Press.
Schimmelfennig F (2016) Good governance and differentiated integration: Graded membership in the European Union. *European Journal of Political Research* **55** (4): 789–810.
Traweek S (1988) *Beamtimes and lifetimes: The world of high energy physicists*. Harvard University Press.
Weinberg A (1961) Impact of large-scale science on the United States. *Science* **134**: 161–4.
Westfall C (2008) Surviving the squeeze: National laboratories in the 1970s and 1980s. *Historical Studies in the Natural Sciences* **38** (4): 475–8.

Index